Collins

AQA GCSE 9-1
Geography

Workbook

Janet Hutson, Dan Major, Paul Berry, Brendan Conway, Tony Grundy, Robert Morris and Iain Palôt

Revision That Really Works

Experts have found that there are two techniques that help you to retain and recall information and consistently produce better results in exams compared to other revision techniques.

It really isn't rocket science either – you simply need to:
- **test yourself** on each topic as many times as possible
- **leave a gap** between the test sessions.

Three Essential Revision Tips

1. **Use Your Time Wisely**
   - Allow yourself plenty of time.
   - Try to start revising at least six months before your exams – it's more effective and less stressful.
   - Don't waste time re-reading the same information over and over again – it's time-consuming and not effective!

2. **Make a Plan**
   - Identify all the topics you need to revise.
   - Plan at least five sessions for each topic.
   - One hour should be ample time to test yourself on the key ideas for a topic.
   - Spread out the practice sessions for each topic – the optimum time to leave between each session is about one month but, if this isn't possible, just make the gaps as big as realistically possible.

3. **Test Yourself**
   - Methods for testing yourself include: quizzes, practice questions, flashcards, past papers, explaining a topic to someone else, etc.
   - Don't worry if you get an answer wrong – provided you check what the correct answer is, you are more likely to get the same or similar questions right in future!

Visit **collins.co.uk/collinsGCSErevision** for more information about the benefits of these techniques, and for further guidance on how to plan ahead and make them work for you.

Command Words used in Exam Questions

This table defines some of the most commonly used command words in GCSE exam questions.

Command word	Meaning
State	Write clearly and plainly
Define	Give the meaning of a term
Calculate	Work out the value of
Outline	Give a brief account or summary
Compare	Identify similarities and differences
Describe	Write what something is or appears to be
Suggest	Propose an idea or solution
Explain	Say why or how
Assess	Make an informed judgement
To what extent	Weigh up the importance or success of
Evaluate	Using evidence, weigh up both sides of an argument
Discuss	Offer key points about the different sides, or strengths and weaknesses, of an issue
Justify	Give detailed reasons for an idea

Contents

The Challenge of Natural Hazards

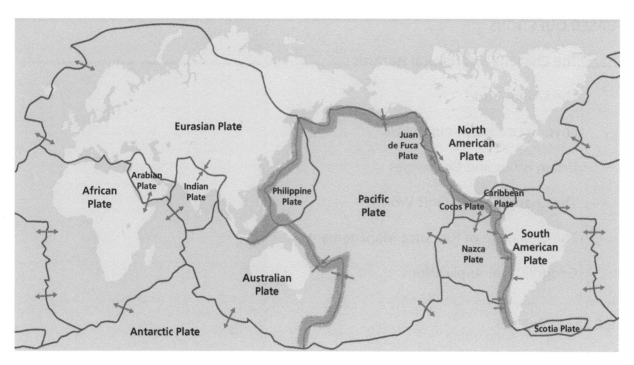

1 **a)** What name is given to the highly active area of tectonic activity shown in orange on the map? [1]

_____

b) Use the map to help you name two plates that form a **destructive** margin. [1]

_____

2 Explain what happens in a **collision zone**. [3]

_____

_____

_____

3 What **three** items would you include in an emergency kit for people living in an earthquake zone? Explain how they might help. [6]

_____

_____

_____

_____

_____

_____

The Challenge of Natural Hazards

4 Why are **tsunamis** so dangerous? [4]

....

....

....

....

5 Describe some of the **longer term problems** which people have to deal with after an earthquake. [8]

....

....

....

....

....

....

....

....

....

6 Describe some of the ways that **volcanic eruptions** can be predicted. [6]

....

....

....

....

....

....

7 **a)** This map shows the tracks of **tropical storms** over the last 70 years.

Describe the pattern of tropical storm tracks in the map. [4]

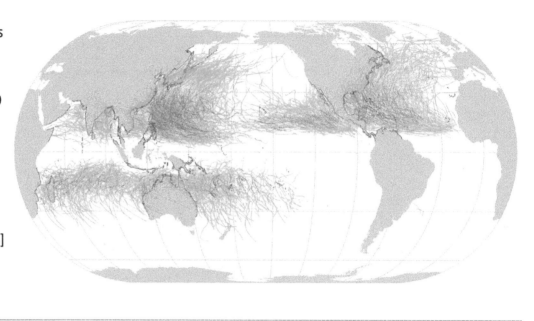

....

....

....

....

b) This map shows **average sea surface temperatures**.

Suggest reasons for the links between the distribution of tropical storms and sea surface temperatures. [4]

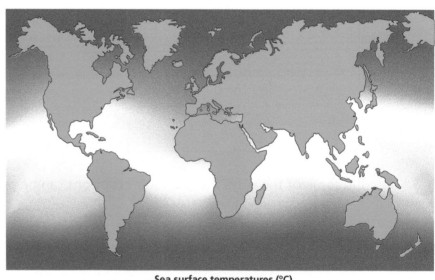

Sea surface temperatures (°C)

0   5   10   15   20   25   30

....

....

....

....

8 The UK has a **maritime** climate. What does this mean? [3]

....

....

....

....

9 What causes the **variations in rainfall** between the west of the UK and those regions in the east? [4]

....

....

....

....

....

10 How are **ice cores** used as evidence of long-term climate change? [3]

....

....

....

....

11 The **impacts of climate change** can be managed through **adaptation and mitigation**.
Explain, using examples, how these two approaches might work. [8]

....

....

....

....

....

....

....

....

Total Marks .... / 55

The Living World

Video Solution Question 6

1 What is **ecotourism**? [2]

2 Large-scale ecosystems are also called '**biomes**'. Name a biome. .... [1]

3 Describe **one** way in which humans can disrupt the **balance of an ecosystem**. [2]

4 Explain how the **nutrient cycle** works. [5]

5 Name the **two** regions where there is **low pressure in January**. [2]

6 What measures allow tropical rainforests to be **managed sustainably**? [6]

The Living World

7 Why do some trees in the tropical rainforest have **buttress roots**? [2]

....

....

....

8 What sort of trees grow in the **boreal forests**? .... [1]

9 How do plants that grow in desert regions **adapt** in **order to survive**? [4]

....

....

....

....

....

....

10 Define the term '**desertification**'. [1]

....

11 a) For **an area you have studied**, explain what has caused the problem of desertification. [4]

....

....

....

....

....

b) Describe what **strategies** have been adopted to deal with desertification in this area. [4]

....

....

....

....

....

The Living World

12 **a)** As the polar regions get warmer, suggest what **resources might be exploited**. [2]

b) Suggest why this exploitation might be a **problem**. [4]

13 What is the **Madrid Protocol** and what does it prohibit? [3]

14 What is **permafrost**? [2]

15 What is **extreme tourism** and what problems might this cause? [4]

Total Marks _____ / 49

Physical Landscapes in the UK

1. What conditions have to exist for **freeze-thaw weathering** to take place? [2]

2. Explain how **abrasion** takes place under a moving glacier. [2]

3. Describe the **shape** and **formation** of **drumlins**. [4]

4. Explain why **glaciated uplands** are unsuitable for farming. [5]

5. Explain what makes glaciated uplands suitable for the **generation of hydroelectric power**. [4]

6 Look at the photograph below showing part of a coast that is being managed.

a) Are the coastal management measures shown, **hard** or **soft** engineering?   [1]

....

b) State the names of the **coastal management structures** shown.   [2]

....

....

7 For a stretch of coastline you have studied, explain why **coastal protection** was necessary and what measures were put in place.   [6]

....

....

....

....

....

....

....

....

8 Describe the conditions necessary for **spit** formation. [3]

....

....

....

....

9 Using an example you have studied, describe the advantages of '**managed retreat**'. [4]

....

....

....

....

....

10 Describe **four** ways by which a river can **erode** its channel. [4]

....

....

....

....

....

11 Explain how **meanders** are formed. [4]

....

....

....

....

....

12 On the diagram of a **storm hydrograph**, label the following: [7]

| Base flow | Rising limb | Lag time | Peak discharge | Falling limb | Storm flow | Peak rainfall |

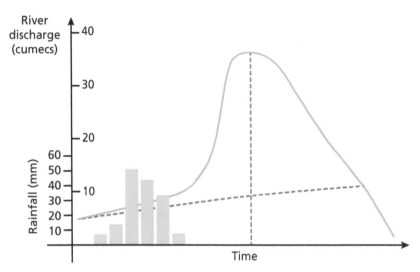

13 For a UK river that you have studied, choose **one** feature caused by **deposition** and describe its location and key features. [4]

_____

_____

_____

_____

14 What are the main causes of **river flooding**? [3]

_____

_____

_____

15 Give **two** advantages and **two** disadvantages of **river hard engineering**. [2]

_____

_____

_____

_____

Total Marks _____ / 57

Urban Issues and Challenges

1 Define the term '**urban sprawl**'. [2]

....

....

2 Suggest why cities in lower income countries (LICs) are growing so rapidly. [3]

....

....

....

3 Suggest **two** reasons why birth rates are higher in LICs. [2]

....

....

4 Give **two** push factors and **two** pull factors which result in rural-to-urban migration. [4]

....

....

....

....

5 a) For a city in a **newly emerging economy** that you have studied, describe **two challenges** it faces. [4]

....

....

....

....

b) Using the same challenges, what **two** solutions has the city implemented? [4]

....

....

....

....

6 Describe the **opportunities** created by **urban change** in a **major UK city** that you have studied. [4]

_____

_____

_____

_____

7 What **challenges** have arisen from **urban change** in a **major UK city** that you have studied? [4]

_____

_____

_____

_____

8 List some of the features of **sustainable living**. [3]

_____

_____

9 Describe, giving examples, **two** strategies that might help **reduce urban traffic congestion**. [4]

_____

_____

_____

_____

10 How does **urban agriculture** help to reduce food miles? [2]

_____

_____

11 Give **two** advantages of **urban greening**. [2]

_____

_____

Total Marks _____ / 38

The Changing Economic World

1

Human Development Index

High human development	0.9–1.00
	0.8–0.89
Medium human development	0.7–0.79
	0.6–0.69
	0.5–0.59
Low human development	0.4–0.49
	0.3–0.39
	0.2–0.29
Not applicable	

Give **two** reasons why a map like this could give a wrong impression.　　　　　　　　　　　[2]

State **three** measures used to show differing rates of development.　　　　　　　　　　　[3]

3 What does the **Demographic Transition Model** help to show?　　　　　　　　　　　[2]

Give **two** physical causes that can lead to **global inequality**.　　　　　　　　　　　[2]

5 What might be **two** of the consequences of **uneven development**?　　　　　　　　　　　[2]

The Changing Economic World

6 What is meant by the term 'intermediate technology'? [2]

....

....

7 Describe a **fair trade** that might help close the **development gap**. [4]

....

....

....

....

8 Describe how **tourism** has reduced the **development gap** in a LIC or NEE that you have studied. [4]

....

....

....

....

9 **TNCs** can play a role in **rapid economic development**. Using an example you have studied, give **two** advantages and **two** disadvantages of rapid economic development. [4]

....

....

....

....

....

10 The provision of aid to a LIC can have advantages and disadvantages. For a country you have studied, state a **type of aid** that has been received and describe the **benefits** to the country. [4]

....

....

....

....

The Changing Economic World

11 What are the main features of the **north-south divide** in the UK? [4]

....

....

....

....

....

12 Define the term **'deindustrialisation'**. [2]

....

....

....

13 Describe measures being taken to make modern industry more **environmentally sustainable**. [4]

....

....

....

....

....

14 What changes might occur in a rural area experiencing **population decline**? [4]

....

....

....

....

....

15 Giving examples, suggest some of the **social** and **economic** benefits from an improvement in the rail infrastructure in the UK. [4]

....

....

....

....

Total Marks _____ / 47

The Challenge of Resource Management

1 Define the term **'carbon footprint'**. [2]

_____

_____

2 Suggest why some of the food eaten in the UK has a high carbon footprint. [3]

_____

_____

_____

3 Suggest reasons for the increase in the **domestic** demand for water in the UK. [2]

_____

_____

4 Explain the difficulties of matching **supply and demand** for water in the UK. [6]

_____

_____

_____

_____

_____

_____

5 Suggest why the UK's reliance on **fossil fuels** is seen as a security risk. [4]

_____

_____

_____

_____

_____

6 Study the figure below. Suggest reasons why some regions of the world have **water stress** or **scarcity**. [4]

Fresh Water in the World
Access to renewable water sources (m³ per person, per year)

Legend:
- No data
- 0 – 1000 (Scarcity)
- 1000 – 2500 (Stress / vulnerability)
- 2500 – 15 000
- 15 000 – 50 000+

....

....

....

....

7 What are **renewables**? [3]

....

....

....

8 Outline why some people have objections to **fracking**. [3]

....

....

....

....

The Challenge of Resource Management

9 Many countries in Africa suffer from **food insecurity**. Suggest **two** reasons for this. [4]

10 Describe what conditions might lead to a **famine**. [4]

11 One way of increasing food supply is by use of irrigation. What is **irrigation**? [1]

12 Suggest how **biotechnology** can help to increase food supply. [3]

13 Using an example you have studied, show how **urban farming** is important in feeding those who are less well off. [3]

14 Water usage is increasing worldwide but some places suffer from water scarcity. Using examples, suggest what can be done to ensure **sustainable water supplies**. [6]

15 Suggest reasons for the increase in **energy consumption** worldwide. [3]

16 Describe how energy usage can be made more **sustainable**. [5]

Total Marks _____ / 56

Geographical Applications

1   For your geographical enquiry, describe and explain patterns in your data and any anomalies which did not correspond to the main patterns.   [8]

_____

_____

_____

_____

_____

_____

_____

_____

_____

_____

_____

_____

_____

_____

_____

Total Marks _____ / 8

Collins

GCSE
GEOGRAPHY
Paper 1 Living with the physical environment

Time allowed: 1 hour 30 minutes

Materials

For this paper you must have:

- a pencil
- a rubber
- a ruler.

You may use a calculator.

Instructions

- Use black ink or a black ball-point pen.
- Answer **all** questions in Section A and Section B.
- Answer **two** questions in Section C.
- Cross through any work you do not want to be marked.

Information

- The marks for questions are shown in brackets.
- The total number of marks available for this paper is 88.
- Spelling, punctuation, grammar and specialist terminology will be assessed in Question 01.10.

Advice

- For the multiple-choice questions, completely fill in the circle alongside the appropriate answer(s).

  CORRECT METHOD [●]　　　　WRONG METHODS [⊗] [⊙] [⊜] [☑]

- If you want to change your answer, you must cross out your original answer as shown. [⊠]

- If you wish to return to an answer previously crossed out, ring the answer you now wish to select as shown. (⊠)

Name: ....

Practice Exam Paper 1

Section A: The challenge of natural hazards

Answer **all** questions in this section.

Question 1   The challenge of natural hazards

Figure 1 shows some of the world's tectonic plates and their direction of movement. Two countries which experienced earthquakes in 2018, the UK and Japan, are indicated.

Figure 1

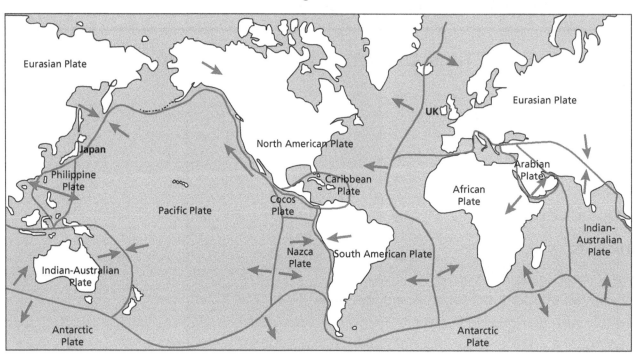

0 1 · 1 Using **Figure 1**, which one of the following statements is true?   **[1 mark]**

Shade **one** circle only.

A   There is a constructive plate margin between the Pacific Plate and the Philippine Plate.   ⬭

B   There is a constructive plate margin between the Pacific Plate and the North American Plate.   ⬭

C   There is a constructive plate margin between the Eurasian Plate and the North American Plate.   ⬭

D   There is a constructive plate margin between the Arabian Plate and the Eurasian Plate.   ⬭

0 1 · 2 Using **Figure 1**, name the type of plate margin between the South American and the Nazca plates. **[1 mark]**

....

0 1 · 3 With the help of **Figure 1**, explain why Japan experiences more earthquakes than the UK. **[2 marks]**

....

....

....

....

Study **Figure 2**, a photograph showing the aftermath of an earthquake in Sapporo, Japan, in September 2018.

Figure 2

Question 1 continues on the next page

0 1 · 4 Suggest how tectonic hazards can have primary and secondary effects.

Use **Figure 2** and your own understanding. **[6 marks]**

_____

_____

_____

_____

_____

_____

_____

_____

_____

_____

Study **Figure 3**, a map showing global air circulation and surface winds.

Figure 3

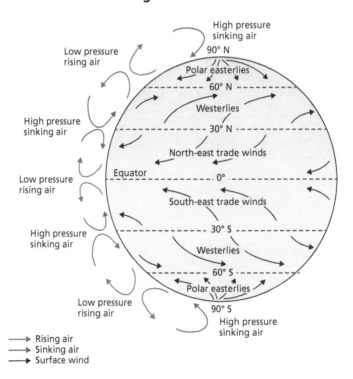

0 1 · 5 Using **Figure 3**, which **two** of the following statements are true? [2 marks]

Shade **two** circles only.

A   Air sinks at the Equator.   ○

B   There is high pressure at the North Pole.   ○

C   There is sinking air in areas of low pressure.   ○

D   The north-east trade winds blow from 30°N towards the Equator.   ○

E   Westerlies blow from the Equator towards 30° North and South.   ○

F   There is low pressure at 30° North and South.   ○

0 1 · 6 Using **Figure 3**, describe the link between air pressure and surface winds.   [1 mark]

....

....

0 1 · 7 Explain the causes of tropical storms.   [4 marks]

....

....

....

....

....

....

....

Question 1 continues on the next page

Study **Figure 4**, a table listing some Atlantic Ocean tropical storms since 2004.

Figure 4

Tropical storm	Number of deaths	Maximum wind speed (miles per hour)
Hurricane Charley, 2004	15	150
Hurricane Katrina, 2005	1200	175
Hurricane Ike, 2008	195	145
Hurricane Igor, 2010	4	155
Hurricane Irene, 2011	49	121
Superstorm Sandy, 2012	285	110
Hurricane Matthew, 2016	603	165
Hurricane Michael, 2018	43	161

0 1 · 8 'As the maximum wind speed of tropical storms increases, so do the number of deaths.'

Do you agree with this statement?

Use evidence from **Figure 4** to support your answer. **[2 marks]**

0 1 · 9 Suggest how climate change might affect the distribution and intensity of tropical storms. **[2 marks]**

Study **Figure 5**, photographs showing strategies used to manage climate change.

Figure 5

Mitigation strategies

Alternative energy production

Planting trees

Adaptation strategies

Reducing risks from rising sea levels

Change in agricultural systems

Question 1 continues on the next page

Practice Exam Paper 1

0 1 . 10 'Managing climate change involves both mitigation (reducing causes) and adaptation (responding to change).'

Do you agree with this statement?

Explain your answer.

Use **Figure 5** and your own understanding.   **[9 marks] [+ 3 SPaG marks]**

_____

_____

_____

_____

_____

_____

_____

_____

_____

_____

_____

_____

_____

_____

_____

_____

_____

End of Section A

Section B: The living world
Answer **all** questions in this section.

Question 2   The living world

Study **Figure 6**, a world map showing the distribution of two types of forest biome.

Figure 6

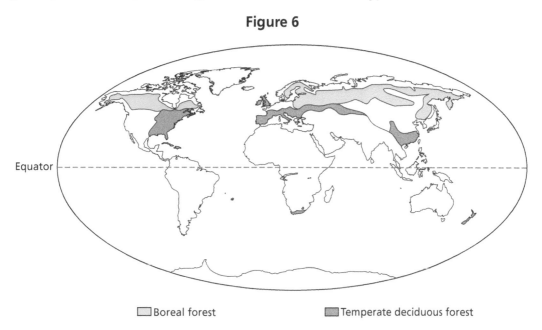

Equator

☐ Boreal forest   ■ Temperate deciduous forest

---

0 2 · 1 Using **Figure 6**, which of the statements below is **true**?   **[1 mark]**

Shade **one** circle only.

A   There is temperate deciduous forest in every continent.   ◯

B   There is no boreal forest in the Southern Hemisphere.   ◯

C   Worldwide, there is more temperate deciduous forest than boreal forest.   ◯

D   Boreal forest is always found on the western side of continents.   ◯

0 2 · 2 Using **Figure 6**, which of the statements below about the climate of temperate deciduous forest is **true**?   **[1 mark]**

Shade the circle next to the correct statement.

A   Short mild summers, long cold winters, often below 0°C.   ◯

B   Warm summers, mild winters, rain falling all year round.   ◯

Question 2 continues on the next page

0 2 · 3 What is an ecosystem?   [1 mark]

....

0 2 · 4 State **one** role of decomposers in an ecosystem.   [1 mark]

....

Study **Figure 7**, which shows part of the food web for an area of Scottish upland.

Figure 7

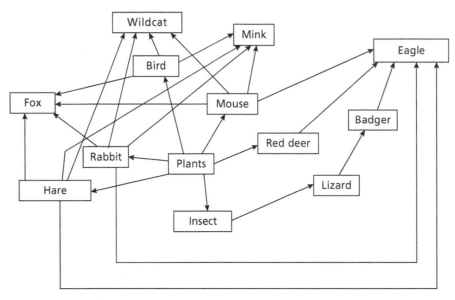

0 2 · 5 Using **Figure 7**, give **one** example of a primary consumer.   [1 mark]

....

0 2 · 6 Suggest what would happen in the food web shown in **Figure** 7 if red deer
became extinct.   [2 marks]

....

....

....

....

Study **Figure 8**, which shows a climate graph for Manaus, a city in the Amazon Rainforest in Brazil.

Figure 8

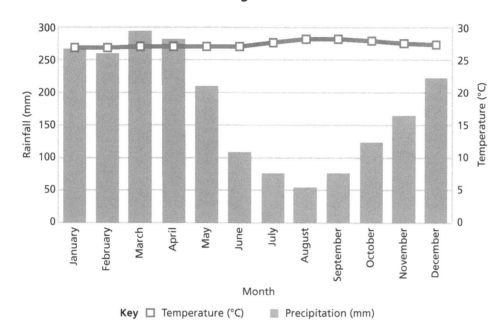

0 2 . 7 Using **Figure 8**, describe the pattern of rainfall. **[2 marks]**

....

....

....

....

0 2 . 8 Give **one** reason why tropical rainforests have high temperatures throughout the year. **[1 mark]**

....

....

Question 2 continues on the next page

Practice Exam Paper 1

Study **Figure 9**, which shows photographs of the Amazon Rainforest in Brazil.

Figure 9

0 2 · 9   Using **Figure 9** and your own understanding, explain how deforestation has economic and environmental impacts. **[6 marks]**

....

....

....

....

....

....

....

....

....

....

0 2 · 10 'Plants and animals adapt to survive in harsh physical conditions.'

Choose **one** of the following environments.

Hot desert environment ☐

Cold environment ☐

Tick the box to show which environment you have chosen.

Using a case study, to what extent do you agree with this statement? **[9 marks]**

....

....

....

....

....

....

....

....

....

....

....

....

....

....

....

....

....

End of Section B

Section C: Physical landscapes in the UK

Answer **two** questions from the following:
Question 3 (Coasts), Question 4 (Rivers), Question 5 (Glacial)

Shade the circle below to indicate which **two** optional questions you will answer.

Question 3   ⬭   Question 4   ⬭   Question 5   ⬭

Question 3   Coastal landscapes in the UK

Study **Figure 10**, a 1: 25 000 Ordnance Survey map extract of an area near to Scarborough in North Yorkshire.

Figure 10

0   1   2   3
km

Scale 1 : 25 000
4 centimetres to 1 kilometre (one grid square)

0 3 · 1 Using **Figure 10**, give the six-figure grid reference for Red Cliff Hole. **[1 mark]**
Shade **one** circle only.

A	073844	◯	B	085837	◯
C	079842	◯	D	099839	◯

0 3 · 2 Using **Figure 10**, which of the following coastal features is **not** shown in grid square 0983? **[1 mark]**

Shade **one** circle only.

A	A beach	◯	B	A wave cut platform	◯
C	A cliff	◯	D	A spit	◯

0 3 · 3 Using **Figure 10**, what is the distance (in metres) between the mean high water and the mean low water at **point A** in grid square 0784? **[1 mark]**

0 3 · 4 Using **Figure 10**, describe **one** piece of evidence which suggests that the area on the map attracts tourists. **[2 marks]**

0 3 · 5 Explain the formation of sand dunes. **[4 marks]**

Question 3 continues on the next page

Study **Figure 11**, photographs showing some hard engineering strategies.

Figure 11

0 3 · 6 Discuss the costs and benefits of hard engineering strategies in protecting coastlines.   **[6 marks]**

Question 4   River landscapes in the UK

Study **Figure 12**, a 1: 25 000 Ordnance Survey map extract of part of the south Pennines in Lancashire.

Figure 12

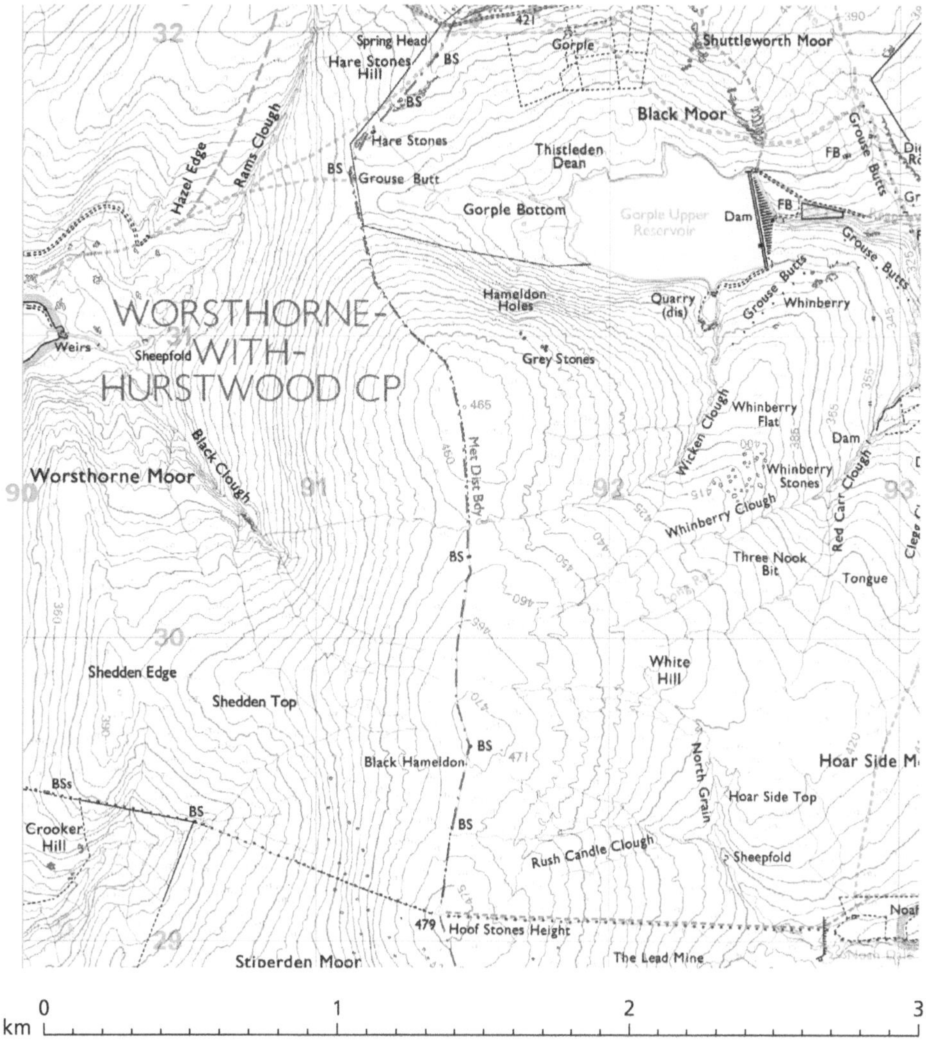

Scale 1 : 25 000
4 centimetres to 1 kilometre (one grid square)

0 4 · 1 In grid square 9131 of **Figure 12**, which area has the steepest slopes?   **[1 mark]**

Shade **one** circle only.

A   South-East   ○

B   North-East   ○

C   South-West   ○

D   North-West   ○

Question 4 continues on the next page

0 4 · 2 Using **Figure 12**, describe the shape of the land around Black Clough river in grid square 9030. **[1 mark]**

....

....

0 4 · 3 Using **Figure 12**, which location is at a confluence? **[1 mark]**

Shade **one** circle only.

A   914293   ○

B   905306   ○

C   920305   ○

D   929311   ○

0 4 · 4 Using **Figure 12** and your own understanding, describe the river erosion processes which will be taking place in this area. **[2 marks]**

....

....

....

....

0 4 · 5 Explain how river levées are formed. **[4 marks]**

....

....

....

....

....

....

....

Collins

GCSE
GEOGRAPHY

Paper 2 Challenges in the human environment

Time allowed: 1 hour 30 minutes

Materials

For this paper you must have:

- a pencil
- a rubber
- a ruler.

You may use a calculator.

Instructions

- Use black ink or a black ball-point pen.
- Answer **all** questions in Section A and Section B.
- Answer Question 3 and **one other** question in Section C.
- Cross through any work you do not want to be marked.

Information

- The marks for questions are shown in brackets.
- The total number of marks available for this paper is 88.
- Spelling, punctuation, grammar and specialist terminology will be assessed in Question 01.10.

Advice

- For the multiple-choice questions, completely fill in the circle alongside the appropriate answer(s).

CORRECT METHOD [●]   WRONG METHODS [⊗] [◉] [⊜] [✓]

- If you want to change your answer, you must cross out your original answer as shown.
- If you wish to return to an answer previously crossed out, ring the answer you now wish to select as shown. ⊗

Name: _____

Practice Exam Paper 2

Section A: Urban issues and challenges

Answer **all** questions in this section.

Question 1 Urban issues and challenges

Study **Figure 1**, a map showing global natural population increase.

Figure 1

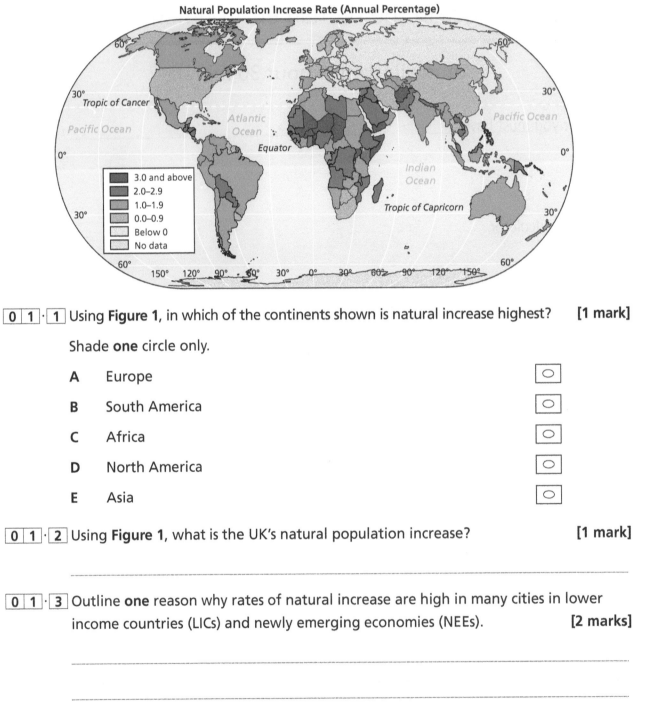

Natural Population Increase Rate (Annual Percentage)

Legend:
- 3.0 and above
- 2.0–2.9
- 1.0–1.9
- 0.0–0.9
- Below 0
- No data

0 1 · 1 Using **Figure 1**, in which of the continents shown is natural increase highest? **[1 mark]**

Shade **one** circle only.

A   Europe   ⬭

B   South America   ⬭

C   Africa   ⬭

D   North America   ⬭

E   Asia   ⬭

0 1 · 2 Using **Figure 1**, what is the UK's natural population increase? **[1 mark]**

....

0 1 · 3 Outline **one** reason why rates of natural increase are high in many cities in lower income countries (LICs) and newly emerging economies (NEEs). **[2 marks]**

....

....

....

Study **Figure 2**, a photograph of a squatter settlement in Rio de Janeiro, Brazil.

Figure 2

0 1 · 4 Using **Figure** 2, suggest **one** problem faced by people in Rio de Janeiro as a result of urban growth. **[2 marks]**

....

....

....

....

0 1 · 5 Using an example, describe how urban planning is improving the quality of life for the urban poor. **[6 marks]**

....

....

....

....

....

....

....

....

Question 1 continues on the next page

0 1 · 6 What is 'urban greening'? [1 mark]

_____

_____

Study **Figure 3**, a photograph of One Central Park in Sydney, Australia.

Figure 3

0 1 · 7 Explain why creating green space is important for sustainable urban living.

Use **Figure 3** and your own understanding. [4 marks]

_____

_____

_____

_____

_____

_____

_____

_____

0 1 · 8 Complete the following fact file for a UK city that you have studied. [2 marks]

Name of UK city	
Location in the UK	
Importance in the UK	

0 1 · 9 Describe the impacts of international migration on the character of a UK city that you have studied. [2 marks]

....

....

....

....

Study **Figure 4**, photographs of Ancoats in Manchester before and after regeneration.

Figure 4

Question 1 continues on the next page

Practice Exam Paper 2

0 1 · 10 Assess the extent to which a regeneration project can solve urban problems.

Use **Figure 4** and an example you have studied.　　　　　**[9 marks] [+ 3 SPaG marks]**

End of Section A

Section B: The changing economic world

Answer **all** questions in this section.

Question 2   The changing economic world

Study **Figure 5**, a table showing Gross National Income (GNI) data for selected countries in 2020.

Figure 5

Name of Country	GNI (US$ per person)
Argentina	9070
Bangladesh	2030
Denmark	63,010
Ethiopia	890
India	1920
Italy	32,290
Kenya	1840
Mali	830
New Zealand	41,550
Portugal	21,790
Saudi Arabia	21,930

`0 2 · 1` Calculate the median value for the GNI data in **Figure 5**.   **[2 marks]**

Median = ...

`0 2 · 2` Give **two** social measures of development.   **[2 marks]**

....

....

Question 2 continues on the next page

Practice Exam Paper 2

0 2 . 3 Outline the limitations of economic measures of development. [3 marks]

....

....

....

....

....

....

0 2 . 4 Explain how fair trade can reduce the development gap. [4 marks]

....

....

....

....

....

....

....

0 2 . 5 Describe the environmental context of a named lower income country (LIC) or newly emerging economy (NEE).

Name of country: .... [2 marks]

....

....

....

....

0 2 · 6 Using a case study of a LIC or NEE, discuss the advantages and disadvantages of transnational corporations to the country.

[6 marks]

....

....

....

....

....

....

....

....

....

....

Study **Figure 6**, pie charts showing how the UK's employment structure has changed over time.

Figure 6

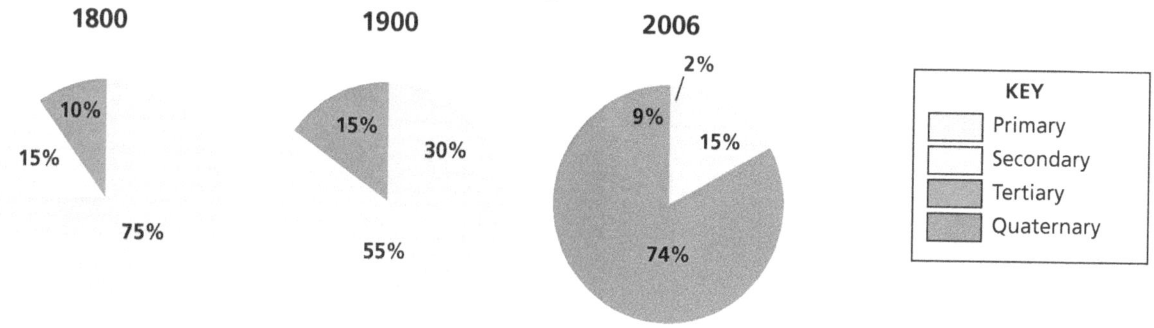

1800: 10%, 15%, 75%
1900: 15%, 30%, 55%
2006: 2%, 9%, 15%, 74%

KEY
Primary
Secondary
Tertiary
Quaternary

0 2 · 7 Using **Figure 6**, which **two** of the following statements are true?

[2 marks]

Shade **two** circles only.

A   In 1800, most people were employed in the secondary sector.   ◯

B   The largest employment sector in 2006 was the tertiary sector.   ◯

C   Employment in the primary sector has decreased over time.   ◯

D   In 1900, most people were employed in the quaternary sector.   ◯

E   Employment in the primary sector has stayed constant over time.   ◯

F   The tertiary sector has always employed the most people.   ◯

Question 2 continues on the next page

Practice Exam Paper 2

0 2 . 8 To what extent have strategies to reduce the UK's north-south divide been successful?

[9 marks]

_____

_____

_____

_____

_____

_____

_____

_____

_____

_____

_____

_____

End of Section B

Section C: The challenge of resource management

Answer Question 3 and **either** Question 4 **or** Question 5 **or** Question 6.

Question 3   The challenge of resource management

Study **Figure 7**, a graph showing the daily caloric intake of selected countries around the world.

Figure 7

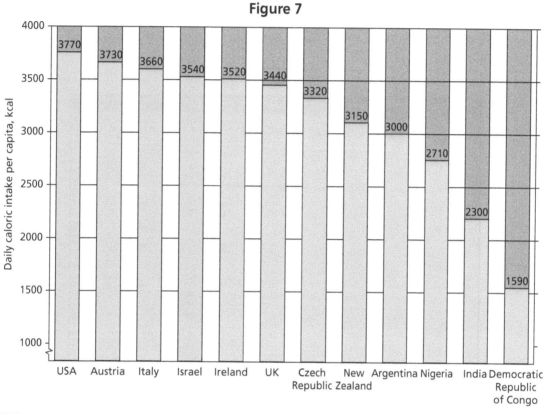

0 3 · 1 Using **Figure 7**, what is the difference in daily caloric intake per capita between the USA and the Democratic Republic of Congo?   **[1 mark]**

0 3 · 2 Using **Figure 7** and your own understanding, suggest how daily caloric intake can influence well-being.   **[3 marks]**

Question 3 continues on the next page

0 3 · 3 Outline **one** disadvantage of importing food from other countries.   **[2 marks]**

....

....

....

Study **Figure 8**, a map showing levels of water stress across England and parts of Wales.

Figure 8

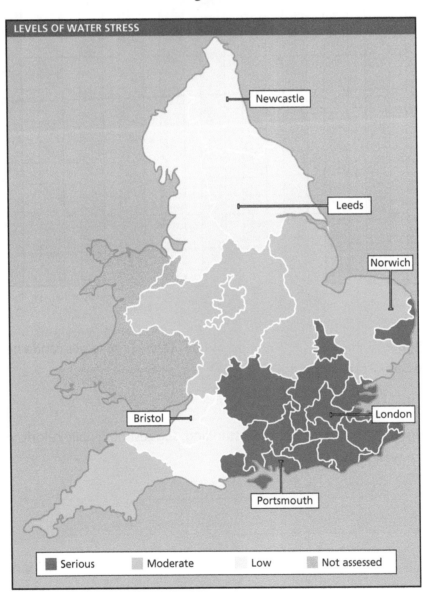

0 3 · 4 Using **Figure 8** and your own understanding, discuss the issues arising from the UK's changing demand for water.

[6 marks]

_____

_____

_____

_____

_____

_____

_____

_____

_____

_____

_____

Section C continues on the next page

Answer **either** Question 4 (Food) **or** Question 5 (Water) **or** Question 6 (Energy).

Question 4   Food

Study **Figure 9**, a map showing the percentage of the population that are undernourished in the continent of Africa.

Figure 9

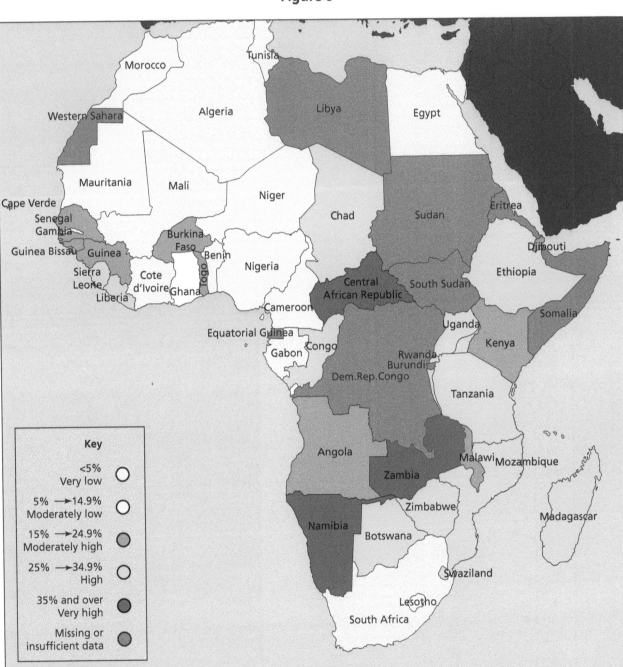

0 4 · 1 Using **Figure 9**, name **one** country where less than 5% of the population is
undernourished.

[1 mark]

0 4 . 2 Using **Figure 9**, how many of the African countries are shown to have over 35% of their population undernourished? **[1 mark]**

Shade **one** circle only.

A   3   ◯

B   5   ◯

C   7   ◯

D   8   ◯

0 4 . 3 Using **Figure 9**, describe the distribution of the countries which had less than 5% of their population undernourished. **[2 marks]**

....

....

....

....

0 4 . 4 Suggest **one** reason for differences in undernourishment between countries. **[2 marks]**

....

....

....

....

0 4 . 5 What is meant by 'food insecurity'? **[1 mark]**

....

....

Question 4 continues on the next page

Study **Figure 10a** and **Figure 10b**, photographs showing strategies to increase food supply.

Figure 10a – Using hydroponics to
grow crops

Figure 10b – Irrigating crops in
Burkina Faso

0 4 · 6 Using **Figure 10a** and **Figure 10b**, explain how countries can increase their
food supply.

[6 marks]

Question 5   Water

Study **Figure 11**, a map showing predicted levels of water stress across Europe in 2040.

Figure 11

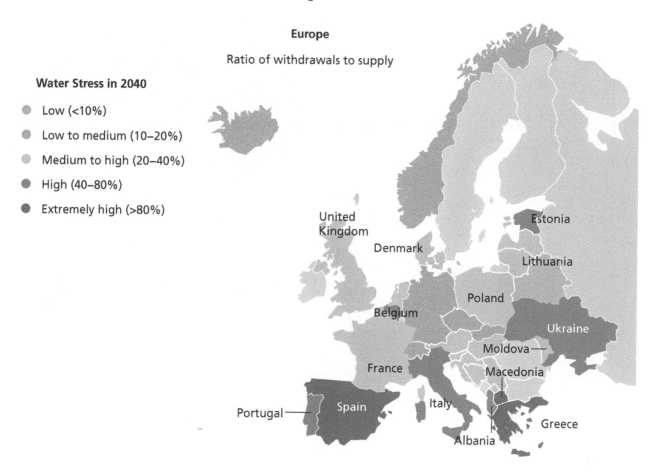

0 5 · 1 Using **Figure 11**, name **one** country with a water stress rating of over 80%.   **[1 mark]**

---

0 5 · 2 Using **Figure 11**, how many European countries have a water stress ratio of less than 10%?   **[1 mark]**

Shade **one** circle only.

A   5

B   7

C   10

D   12

Question 5 continues on the next page

Practice Exam Paper 2

0 5 . 3 Using **Figure 11**, describe the distribution of the countries which have a water stress ratio of less than 10%.   **[2 marks]**

....

....

....

....

0 5 . 4 Suggest **one** reason for differences in water stress between countries.   **[2 marks]**

....

....

....

....

0 5 . 5 What is meant by 'water insecurity'?   **[1 mark]**

....

....

Study **Figure 12a** and **Figure 12b**, photographs showing strategies to increase water supply.

Figure 12a – A Desalination Plant at Arrecife, Lanzarote, Canary Islands

Figure 12b – Ridgegate Reservoir near Macclesfield in Cheshire, UK

`0 5 · 6` Using **Figure 12a** and **Figure 12b**, explain how higher income countries can increase their water supply. **[6 marks]**

....

....

....

....

....

....

....

....

....

....

....

Question 6 Energy

Study **Figure 13**, a map showing projected global change in primary energy demand.

Figure 13

Change in primary energy demand, 2016–40 (Million tonnes of oil equivalent)

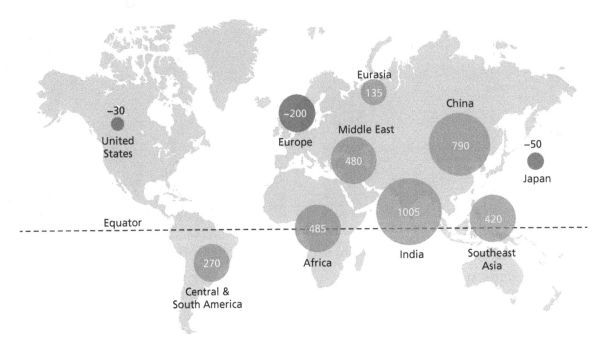

0 6 · 1 Using **Figure 13**, which part of the world is projected to experience the highest
increase in energy demand? **[1 mark]**

0 6 · 2 Using **Figure 13**, how many parts of the world are predicted to see a fall in energy demand? [1 mark]

Shade **one** circle only.

A   2   ○

B   3   ○

C   4   ○

D   5   ○

0 6 · 3 Using **Figure 13**, describe the distribution of the parts of the world which are predicted to see a fall in energy demand. [2 marks]

....

....

....

....

0 6 · 4 Suggest **one** reason for differences in predicted energy demand changes between different countries. [2 marks]

....

....

....

....

0 6 · 5 What is meant by 'energy insecurity'? [1 mark]

....

....

Question 6 continues on the next page

Study **Figure 14a** and **Figure 14b**, photographs showing strategies to increase energy supply.

Figure 14a – North Hoyle Wind Farm off the coast of North Wales

Figure 14b – A nuclear power plant near Thionville in France

0 6 · 6 Using **Figure 14a** and **Figure 14b**, explain how countries can increase their energy supply.

[6 marks]

....

....

....

....

....

....

....

....

....

....

....

End of Section C

END OF QUESTIONS

Collins

GCSE
GEOGRAPHY
Paper 3 Geographical applications

Time allowed: 1 hour 15 minutes

Materials

For this paper you must have:

- a clean copy of the pre-release resources booklet (see page 81).
- a pencil
- a rubber
- a ruler.

You may use a calculator.

Instructions

- Use black ink or a black ball-point pen.
- Answer **all** questions.
- Cross through any work you do not want to be marked.

Information

- The marks for questions are shown in brackets.
- The total number of marks available for this paper is 76.
- Spelling, punctuation, grammar and specialist terminology will be assessed in Questions 03.3 and 05.4

Advice

- For the multiple-choice questions, completely fill in the circle alongside the appropriate answer(s).

CORRECT METHOD ⬤   WRONG METHODS ⊗ ⊙ ⊜ ✓

- If you want to change your answer, you must cross out your original answer as shown. ⊠

- If you wish to return to an answer previously crossed out, ring the answer you now wish to select as shown. ⊘

Name: ....

Practice Exam Paper 3

Section A: Issue evaluation

Answer **all** questions in this section.

Study **Figure 1**, a graph showing the sources of energy used to create electricity in the UK between 1996 and 2017.

Figure 1

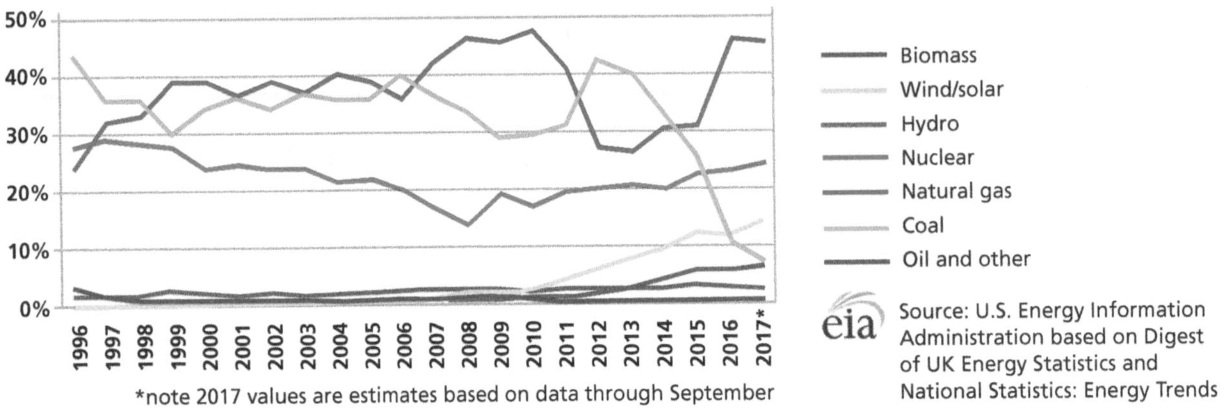

50%
40%
30%
20%
10%
0%
1996
1997
1998
1999
2000
2001
2002
2003
2004
2005
2006
2007
2008
2009
2010
2011
2012
2013
2014
2015
2016
2017*
*note 2017 values are estimates based on data through September
Biomass
Wind/solar
Hydro
Nuclear
Natural gas
Coal
Oil and other
eia
Source: U.S. Energy Information Administration based on Digest of UK Energy Statistics and National Statistics: Energy Trends

0 1 · 1 Using **Figure 1**, in which year did coal and nuclear provide the same amount of electricity? **[1 mark]**

Shade **one** circle only.

A	2015	○	**B**	2017	○
C	2000	○	**D**	2008	○

0 1 · 2 What is meant by 'fossil fuel'? **[1 mark]**

....

....

Study **Figure 2**, a chart which shows the origin of gas supplies in the UK in 2016.

Figure 2

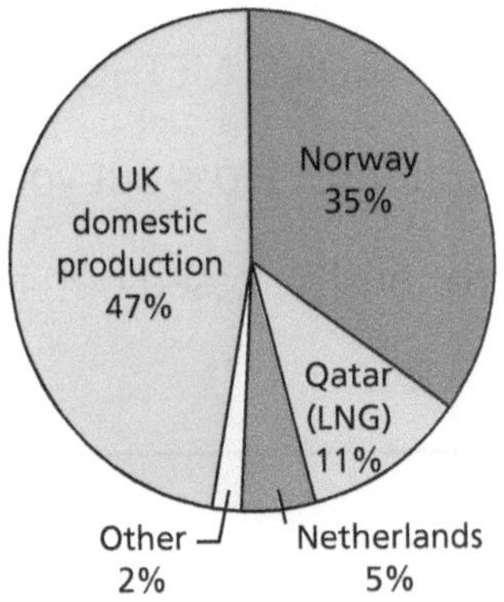

Norway 35%
UK domestic production 47%
Qatar (LNG) 11%
Other 2%
Netherlands 5%

0 1 . 3 Using **Figure 2**, describe the pattern shown on the chart. [2 marks]

....

....

....

....

0 1 . 4 Outline **one** disadvantage of importing energy from other countries. [2 marks]

....

....

....

....

0 1 . 5 'Economic and environmental issues are caused by the exploitation of energy sources.'

Discuss this statement. [6 marks]

....

....

....

....

....

....

....

....

....

....

....

....

Section A continues on the next page

`0 2 · 1` What is meant by 'renewable energy'?   [2 marks]

....

....

`0 2 · 2` Suggest reasons for the growing significance of renewable energy sources in
the UK.   [6 marks]

....

....

....

....

....

....

....

....

....

....

....

Study **Figure 3**, a table showing the changes in UK wind power capacity, generation and percentage of total electricity provided by wind power.

Figure 3

Year	Capacity (MW)	Generation (GW/h)	% of total electricity use
2011	6,540	12,675	3.81
2012	8,871	20,710	5.52
2013	10,976	24,500	7.39
2014	12,440	28,100	9.30
2015	13,602	40,442	11.0
2016	16,218	37,368	12.0
2017	19,837	49,607	17.0
2018	21,606	56,907	17.1
2019	23,882	63,795	19.7
2020	24,485	75,369	24.1

`0 3 · 1` Using **Figure 3**, what was the increase in wind power as a percentage of total electricity use between 2011 and 2020? **[1 mark]**

Shade **one** circle only.

A   24.1   ⬭

B   3.81   ⬭

C   27.91   ⬭

D   20.29   ⬭

Study **Figure 4**, a photograph of a wind farm in the Scottish Highlands.

Figure 4

`0 3 · 2` Using **Figure 4** and your own understanding, suggest reasons why the UK has got good potential for developing wind energy. **[4 marks]**

....

....

....

....

....

....

....

Section A continues on the next page

Practice Exam Paper 3

0 3 . 3 'Developing wind energy projects in the Scottish Highlands are a good idea.'

Do you agree with this statement?   **[9 marks] [+ 3 SPaG marks]**

Yes ☐   No ☐

Tick the box to show your choice.

Use evidence from the resources booklet and your own understanding to explain your answer.

....

....

....

....

....

....

....

....

....

....

....

....

....

End of Section A

Section B: Fieldwork
Answer **all** questions in this section.

A group of students visit part of Snowdonia in North Wales to carry out a fieldwork enquiry. They want to investigate the hypothesis that 'the size of pebbles decreases in size as distance down the valley increases'.

In order to do this, the students measure the long axis (longest side) of stones in material deposited by glaciers. They take the measurement of 100 stones from each of five evenly spaced locations shown in **Figure 5** below.

Figure 5

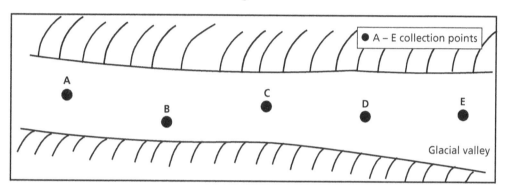

Glacial valley

| 0 4 | · | 1 | Identify the type of sampling method used in **Figure 5**. | [1 mark] |

Shade **one** circle only.

A   Random sampling   ◯

B   Stratified sampling   ◯

C   Systematic sampling   ◯

D   Opportunity sampling   ◯

| 0 4 | · | 2 | Suggest why the type of sampling shown in **Figure 5** is not always possible in a fieldwork enquiry. | [2 marks] |

....

....

....

....

Section B continues on the next page

Figure 6 shows the mean length of stones, in each of the five locations.

Figure 6

	Location A	Location B	Location C	Location D	Location E
Distance down the valley	500 m	1000 m	1500 m	2000 m	2500 m
Mean	72 mm	94 mm	103 mm	166 mm	208 mm

0 4 · 3 Suggest a suitable method for presenting the data shown in **Figure 6**.

Give a reason for your choice. **[2 marks]**

....

....

....

....

0 4 · 4 Suggest **one** advantage and **one** disadvantage of using the mean as a measure
of central tendency. **[2 marks]**

Advantage

....

....

Disadvantage

....

....

0 4 · 5 Outline the conclusions that the students could make from the data in **Figure 6**. **[2 marks]**

....

....

....

....

On a different day, the students wanted to investigate the quality of two footpaths.
Students asked 50 people their opinion of the quality of two footpaths.

Study **Figure 7**, a table showing the results of their survey.

Figure 7

Quality of footpath	Footpath A	Footpath B
Very good	25	10
Good	15	18
Fair	8	12
Poor	2	10

0 4 · 6 Complete **Figure 8** below to show the survey results for Footpath B. **[1 mark]**

Figure 8

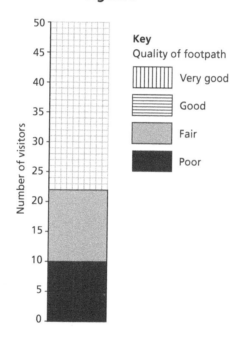

0 4 · 7 What percentage of visitors thought the quality of Footpath B was good or
very good? **[1 mark]**

....

0 4 · 8 Suggest **one** advantage of using a bar chart to present the data shown in
Figure 7. **[1 mark]**

....

....

Section B continues on the next page

Figure 9 shows the change in width along the first 1000 metres of footpaths A and B. The footpath routes have been made into straight lines to make it easier to interpret the results.

Figure 9

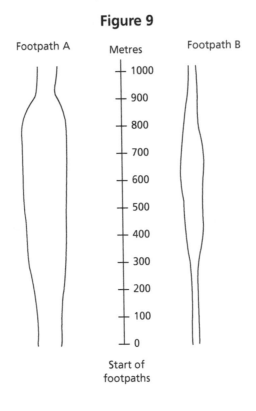

0 4 · 9   Using **Figure 9**, compare the width of the two footpaths.   [4 marks]

....

....

....

....

....

....

....

....

....

Write the title of your physical geography fieldwork enquiry.

Title of physical fieldwork enquiry: ....

....

0 5 · 1 State **one** potential risk you identified in your physical geography fieldwork and **explain** the measure(s) you used to reduce it. **[3 marks]**

Risk

....

Measure(s) used to reduce risk

....

....

....

....

0 5 · 2 Assess the effectiveness of your data collection method(s). **[6 marks]**

....

....

....

....

....

....

....

....

....

....

....

Section B continues on the next page

Write the title of your human geography fieldwork enquiry.

Title of human fieldwork enquiry: ....

....

0 5 · 3 Explain why the chosen location was suitable for the collection of data. **[2 marks]**

....

....

....

0 5 · 4 For **one** of your fieldwork enquiries, to what extent did the data collected allow you
to reach valid conclusions? **[9 marks] [+ 3 SPaG marks]**

Title of fieldwork enquiry: ....

....

....

....

....

....

....

....

....

....

....

....

....

End of Section B

END OF QUESTIONS

Collins

GCSE
GEOGRAPHY

Resources for Paper 3 Geographical applications

Study the resources in this booklet before completing Paper 3.

The resources for Paper 3 of your actual exam will be issued to you 12 weeks before the date of the exam.

Resource 1: Sources of Energy used to Create Electricity in the UK

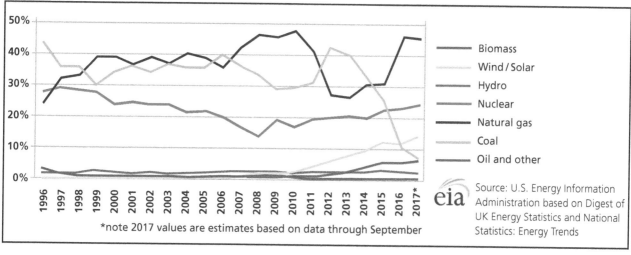

*note 2017 values are estimates based on data through September

Source: U.S. Energy Information Administration based on Digest of UK Energy Statistics and National Statistics: Energy Trends

Resource 2: UK Gas Supplies

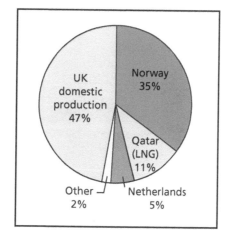

Resources for Paper 3

Resource 3: Issue with Power Generation in January 2017

In January 2017, the UK grid faced a perfect storm stemming from nuclear closures in France, nuclear faults in the UK, a broken inter-connector with France and then on 16th January the wind died for a week. These were the exact conditions that were expected to increase the blackout risk, but the lights stayed on.

Concern about the UK grid is borne out of the closure of 17.7 GW of coal-fired power between 2004 and 2016 and its replacement with 14.4 GW of wind and 10.7 GW of solar.

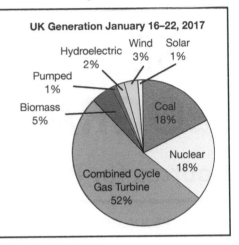

UK Generation January 16–22, 2017

Hydroelectric 2%
Wind 3%
Solar 1%
Pumped 1%
Biomass 5%
Coal 18%
Nuclear 18%
Combined Cycle Gas Turbine 52%

Resource 4: Extract from UK Government Statement, April 2021

The UK government will set the world's most ambitious climate change target into law to reduce emissions by 78% by 2035 compared to 1990 levels.

In line with the recommendation from the independent Climate Change Committee, this sixth Carbon Budget limits the volume of greenhouse gases emitted over a 5-year period from 2033 to 2037, taking the UK more than three-quarters of the way to reaching net zero by 2050. The Carbon Budget will ensure Britain remains on track to end its contribution to climate change while remaining consistent with the Paris Agreement temperature goal to limit global warming to well below 2°C and pursue efforts towards 1.5°C.

Resource 5: Photographs of a Wind Farm in the Scottish Highlands

Resource 6: Contribution of Wind Power to Electricity Generation over Time

Year	Capacity (MW)	Generation (GW/h)	% of total electricity use
2011	6,540	12,675	3.81
2012	8,871	20,710	5.52
2013	10,976	24,500	7.39
2014	12,440	28,100	9.30
2015	13,602	40,442	11.0
2016	16,218	37,368	12.0
2017	19,837	49,607	17.0
2018	21,606	56,907	17.1
2019	23,882	63,795	19.7
2020	24,485	75,369	24.1

Resource 7: Taken from a Statement by the John Muir Trust* in May 2018

The news came at the end of April 2018 that Scottish Ministers had refused consent for two major wind farms that the Trust had campaigned against.

Wild Land Area 34: Reay–Cassley, popular walking country and an area of outstandingly significant geology, was threatened by not just one but three industrial-scale wind farms that could have seen a total of 65 turbines radically change this prized landscape.

To our great relief, the Minister rejected two of these concluding that they were simply inappropriate for such an area.

But the third development, which had come into the planning system later, remained a threat until just a few weeks ago when the Minister published his decision following a Public Local Inquiry into this case in 2017, concluding that "*the proposal would not preserve natural beauty … weighing up all of the material considerations, my conclusion is that, on balance, the adverse consequences of the proposal are too significant to be outweighed by its benefits.*"

Further south the hilly, moorland country of Wild Land Area 19 has been under threat, from a 13-turbine development that would, according to Scottish Natural Heritage, have led to the loss of seven square kilometres of wild land in the north west corner of the Wild Land Area.

This Wild Land Area is also popular walking country, with five Munros and four Corbetts. It has fantastic high mountains and famous cliffs for winter climbing, and a small overlap with the Cairngorms National Park.

Repeated surveys have demonstrated strong public support for wild land protection.

John Muir Trust is a charity founded in 1983 whose mission is "to conserve and protect wild places with their indigenous animals, plants and soils for the benefit of present and future generations".

Resource 8: Taken from a Statement from the RSPB
(*The Royal Society for the Protection of Birds*)

How do wind farms affect birds?

The available evidence suggests wind farms can harm birds in three possible ways – disturbance, habitat loss and collision.

Some poorly sited wind farms have caused major bird casualties, particularly at Tarifa and Navarra in Spain and the Altamont Pass in California. At these sites, planners failed to consider adequately the likely impact of putting hundreds, or even thousands, of turbines in areas which are important for birds of prey.

If wind farms are located away from major migration routes and important feeding, breeding and roosting areas of those bird species known or suspected to be at risk, it is likely they will have minimal impacts.

Reducing the impact on birds

We are involved in scrutinising hundreds of wind farm applications every year to determine their likely wildlife impacts, and we ultimately object to about 6 per cent of those we engage with, because they threaten bird populations. Where developers are willing to adapt plans to reduce impacts to acceptable levels we withdraw our objections; in other cases we robustly oppose them.

However, there are gaps in knowledge and understanding of the impacts of wind energy, so the environmental impact of operational wind farms needs to be monitored - and policies and practices need to be adaptable, as we learn more about the impacts of wind farms on birds.

Resource 9: Financial Benefits to a Local Community Following Wind Farm Construction

Community Benefit Funds
Funds established by wind farm development companies

One example from the Scottish Highlands

In 2014/15, funds worth approximately £1.5 million were managed.

Some donations in 2016:

- Village Hall: the sum of £7000 to provide free access for local groups in 2016 (following a £3000 donation for 2015)
- The Community Plan Group: £6800 for public access automated defibrillators; these are located at St Paul's Church Hall and four other locations
- Baby and Toddler Group: a total of £81 500 for a new play park across from school woods
- Village Summer Activities: the sum of £1409 to help with costs for organising and running the week-long summer school session in the village hall
- £110 towards local events in 2016
- Community Newsletter: a total of £8350 to cover costs
- The Farmers' Association Vintage Rally and Display: the sum of £3429 for 2016
- Shinty Club
- Community Woodlands
- The Music Initiative: £15 440 to cover the costs for two terms of music classes
- Indoor Bowls Club: £1000
- Community Hall: a grant of £1600 to cover the costs of new goalposts and for playing fields grass-cutting
- Seniors' Lunch Club: the sum of £2500
- Primary schools and nurseries: £24 000 for extracurricular activities during the 2015 school year
- Village Growers: the sum of £7800 towards legal and start-up costs for their community garden project
- Care Group: a total of £9600 in 2016
- Community Access and Transport Association: £11 971 towards a new 9-seater bus with wheelchair access
- Reimbursement of ongoing travel costs for those who qualify for free transport

Resource 10: Energy Security

Energy security is the uninterrupted availability of energy sources at an affordable price. In the long term, it attempts to make sure that energy keeps pace with both economic developments and sustainability needs. In the short term, it focuses on the ability of the energy system to react effectively to sudden changes in supply or demand. Lack of energy security has negative economic and social impacts.

Resource 11: UK Wind Power Capability

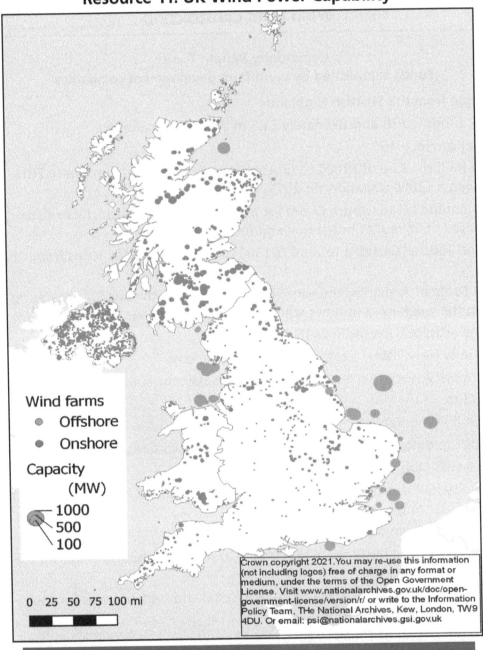

Wind farms
- Offshore
- Onshore

Capacity (MW)
- 1000
- 500
- 100

0   25   50   75   100 mi

Crown copyright 2021. You may re-use this information (not including logos) free of charge in any format or medium, under the terms of the Open Government License. Visit www.nationalarchives.gov.uk/doc/open-government-license/version/r/ or write to the Information Policy Team, THe National Archives, Kew, London, TW9 4DU. Or email: psi@nationalarchives.gsi.gov.uk

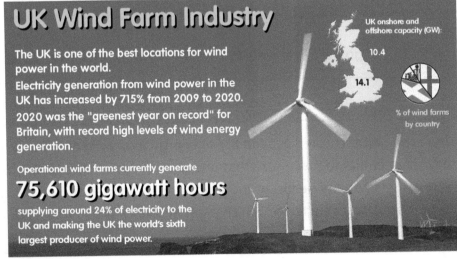

UK Wind Farm Industry

The UK is one of the best locations for wind power in the world.

Electricity generation from wind power in the UK has increased by 715% from 2009 to 2020.

2020 was the "greenest year on record" for Britain, with record high levels of wind energy generation.

Operational wind farms currently generate

75,610 gigawatt hours

supplying around 24% of electricity to the UK and making the UK the world's sixth largest producer of wind power.

UK onshore and offshore capacity (GW):

10.4

14.1

% of wind farms by country

Answers

TOPIC-BASED QUESTIONS

Pages 4–7: The Challenge of Natural Hazards

1   a) Pacific Ring of Fire **[1]**
    b) **Examples:** Pacific Plate and North American Plate; Nazca Plate and South American Plate **[1]**

2   **Example points:** continental plates collide **[1]**; plates buckle rather than subduct **[1]**; fold mountains are formed **[1]**, e.g. Alps, Himalayas **[1]**; no volcanoes **[1]**, but earthquakes occur **[1]**

3   **Example items:** water **[1]**; purifying tablets **[1]**; tinned food **[1]**; first aid kit **[1]**; survival blanket **[1]**; wind-up radio **[1]**; wind-up torch **[1]**; batteries **[1]**. The three named items must be explained as to their usefulness. **[3]**

4   **Example points:** extreme flooding **[1]**; difficult to predict **[1]**; loss of life **[1]**; damage to coastal settlements and infrastructure **[1]**; impacts can be at distance from earthquake (e.g. across the Pacific Ocean) **[1]**; speed of the event **[1]**

5   **Example points:** homelessness **[1]**; buildings destroyed **[1]**; unemployment **[1]**; disruption to transport **[1]**; fatalities **[1]**; shock and trauma **[1]**; PTSD **[1]**; out-migration **[1]**; lack of medical care **[1]**; lack of education **[1]**; rebuilding **[1]**; need for foreign aid **[1]**

6   **Example points:** magma rising causes land surface to swell **[1]**; using tiltmeters **[1]**; measuring chemical composition of gases being emitted **[1]**; monitoring changes in groundwater temperatures **[1]**; monitoring earthquake activity **[1]** with seismometers **[1]**; satellite and aircraft observations **[1]**

7   a) Any suitable answer making the key point that the storms are not inland or along the Equator, but over warm ocean water between 5° and 20° north or south.
      Pacific Ocean: occur in the east Pacific but mainly in the west Pacific near the Philippines and north-east Australia; north Atlantic; Indian Ocean (north and south of the Equator); rare in south Atlantic **[4]**
    b) Areas with higher sea surface temperatures tend to have the most tropical storms **[1]** but this is not the case along the Equator **[1]**. They do not form over land **[1]** and rapidly lose intensity over land because they lose their supply of warm, moist air. **[1]**

8   **Example points:** affected by the sea **[1]**; cool, wet winters **[1]**; warm, dry summers **[1]**

9   **Example points:** high ground in the west **[1]** causes heavier relief rain **[1]** while the east lies in a rain shadow **[1]**, receiving much less rain **[1]**

10 **Example points:** proxy measures **[1]**; measure amount of carbon dioxide and other gases **[1]**, trapped in the ice; pollen analysis **[1]**; show 'ice ages' and interglacials **[1]**

11 **Example points:**
      Adaptation – adjusting to changes in the environment **[1]**; building houses on stilts **[1]**; new agricultural methods **[1]**; building sea defences **[1]**
      Mitigation – reducing greenhouse gas emissions **[1]** and therefore the causes of climate change **[1]**; carbon capture and storage technology **[1]**; planting trees **[1]**; switching to renewable energy **[1]**; international agreements **[1]** such as the Paris Agreement **[1]**

Pages 8–10: The Living World

1   Tourism that uses the beauty of the environment itself as the sole attraction **[1]** and helps to ensure the protection of the environment **[1]**.

2   **Any answer from:** tropical rainforest; tropical grasslands (savanna); temperate grasslands; temperate forest; boreal forest (taiga); tundra; desert **[1]**

3   **Example points:** deforestation **[1]** destroys the habitats of countless organisms, large and small, creating imbalance **[1]**; draining wetlands **[1]**; cutting down hedgerows **[1]**; planting non-native species **[1]**

4   **Example points:** plants grow **[1]**; eaten by herbivores **[1]**; they in turn are eaten by carnivores **[1]**, which die and decompose **[1]** and the remains are reabsorbed by plants **[1]**

5   The Poles **[1]** and the ITCZ (Doldrums, Equator) **[1]**

6   **Example points:** selective logging **[1]**; international agreements such as CITES **[1]**; labelling schemes (FSC) **[1]**; debt reduction **[1]**; encouragement of ecotourism **[1]**; promoting employment **[1]**; reducing out-migration **[1]**

7   To provide stability for the very tall trees **[1]** and to absorb nutrients more easily **[1]**

8   Evergreen coniferous trees **[1]**

9   **Example points:** No leaves **[1]** to reduce transpiration **[1]** and have needles instead **[1]**; widespread root system **[1]**; fleshy leaves and stems **[1]** to retain moisture **[1]**

10 **Example points:** land turning to desert **[1]**; land degradation in dry areas **[1]**

11 a) A named area must be given. Points may include: increasing temperatures **[1]** and reduced rainfall **[1]**, as a result of climate change **[1]**; increased demand for water as a result of increasing population growth **[1]**; overgrazing by cattle **[1]**; excessive irrigation of crops **[1]**
    b) Answer must relate to area studied. Points may include: drip irrigation methods **[1]** and other named appropriate technologies **[1]**; increased recycling of water **[1]**; greater use of greywater **[1]**; limiting the extent of tourist facilities, including golf courses, water parks, etc. **[1]**; tree planting **[1]** because the roots bind the soil together **[1]** and help to stop the soil being blown away **[1]**

12 a) **Example points:** greater exploitation of marine resources **[1]**; fossil fuels extraction **[1]**; along with other minerals **[1]**
    b) **Example points:** contravention of existing protocols **[1]**, such as Madrid **[1]**; overfishing **[1]**; marine pollution **[1]**; land pollution (tar sands) **[1]**; geopolitical tensions **[1]**

13 **Example points:** the Madrid Protocol is part of the Antarctic Treaty System, signed in 1991 **[1]** to protect the environment of Antarctica **[1]**; preserve it as a 'natural reserve, devoted to peace and science' **[1]**; expressly prohibits mining **[1]**

14 **Example points:** permafrost is ground that has been frozen for two or more consecutive years **[1]**; it consists of soil, gravel and sand, usually bound together by ice **[1]**

15 **Example points:** create attraction of hostile environments to the tourist **[1]**; includes polar regions, high mountains, deserts **[1]**; damage to fragile environments and habitats **[1]**; litter and pollution **[1]**; too many tourists in a single area **[1]**; difficulties of international management and agreements **[1]**

Pages 11–14: Physical Landscapes in the UK

1   **Example points:** water must be present **[1]**; temperatures need to go above and below 0°C **[1]**; rocks need to be weakened or cracked **[1]**

2   **Example points:** rock fragments present **[1]**; as the ice moves, these fragments scratch the surface of the solid rock **[1]**, wearing it away **[1]** or leaving scratch marks or striations **[1]**

3   **Example points:** rounded **[1]**, elongated mounds of moraine **[1]** shaped by the moving ice **[1]**. The blunt end faces the approach of the ice **[1]** and the tail points in the direction of the ice as it flowed over **[1]**

4   **Example points:** high **[1]**, steep slopes **[1]**; often impermeable rocks **[1]**; waterlogged soils **[1]**; short growing season **[1]**; windy **[1]**; short daylight hours **[1]**

5   **Example points:** impermeable rocks **[1]**; high rainfall totals **[1]**; steep slopes **[1]**; low value land **[1]**; fast-flowing streams **[1]**; lakes for reservoirs **[1]**

6   a) Hard **[1]**
    b) Rock armour / rip-rap **[1]**; sea wall **[1]**

7   A named stretch of coast must be given. Depending on the chosen location, points may include: protection of settlements **[1]** or infrastructure **[1]**; sea walls **[1]**; rock armour **[1]**; gabions **[1]**; revetments **[1]**; groynes **[1]**; beach feeding **[1]**

8   **Example points:** longshore drift **[1]**; supply of sediment **[1]**; coast changing direction **[1]**, e.g. estuary mouth **[1]**; onshore winds **[1]** from two directions to create curved end **[1]**

9   A named location must be given. Depending on the chosen location, points may include: relatively cheap **[1]**; cost effective **[1]**; creates saltmarsh **[1]**; wildlife habitat **[1]**; good natural defence for low-lying areas **[1]**; restores natural sediment movement **[1]**

10 **Example points:** hydraulic action **[1]**, undermining river banks **[1]**; abrasion **[1]**: rock fragments in water wear away at the bed and banks **[1]**; solution **[1]**: limestone rocks dissolving in slightly

acidic stream water [1]; attrition [1]: particles hitting each other and reducing in size and becoming rounder [1]

11 **Example points:** sinuous route of water in channel (thalweg) [1]; faster water hits outside of bend [1] and erodes more rapidly [1]; river cliffs form [1]; inside bend slower water deposition [1] forms slip-off slope [1]; have an asymmetrical cross section [1]

12

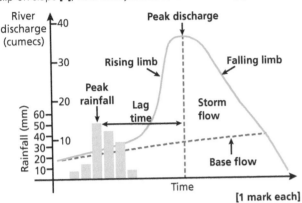

[1 mark each]

13 A named river must be given. Depending on the chosen location, features may include: ox-bow lake, floodplain, levée [4]

14 **Example points:** exceptionally high levels of rainfall [1], in a short period of time [1]; urbanisation brings creation of impermeable surfaces [1], preventing infiltration [1]; upstream changes in land use [1]; deforestation [1]; rapid snow melt [1]; storm surges [1]; exceptionally high tides [1]

15 **Example points:**
Advantages: protection of housing [1], infrastructure [1], railways [1] and electricity supply [1]; can be a tourist attraction [1]; benefits outweigh the costs [1]
Disadvantages: costly to build [1] and maintain [1]; reduces access to the river [1]; traps water should it spill over [1]; unsightly [1]; long-term project [1]; may move flooding problems elsewhere [1]

Pages 15–16: Urban Issues and Challenges

1 Where urban areas grow outwards [1] into the surrounding rural regions [1]

2 **Example points:** rural-to-urban migration [1]; high natural increase [1]; high birth rates and falling death rates [1]; better opportunities [1]

3 **Example points:** high infant mortality rates [1]; lack of medical services [1]

4 **Example points:**
Push: un/underemployment [1]; natural disasters [1]; low wages [1]; difficult farming conditions [1]; lack of alternative job opportunities [1]; isolation [1]; lack of social amenities [1]; lack of schools, hospitals, etc. [1]
Pull: perception of [1] job opportunities [1]; better wages [1]; access to schools, hospitals, etc. [1]; access to transport [1]; better housing conditions [1]; better access to water and electricity [1]; more social amenities [1]

5 a) Specific challenges will depend on the city chosen. Points may include: in-migration from rural areas [1]; informal housing [1]; insufficient healthcare [1]; insufficient education [1]; lack of clean water supply [1]; shortage of electricity [1]; crime [1]; poverty [1]; un/underemployment [1]; pollution [1]; waste disposal [1]

b) Specific solutions will depend on the city chosen and must match with the challenges given in part a). Points may include: upgrading of informal housing [1]; self-build schemes [1]; water and electricity provision [1]; legal rights to land ownership [1]; improved transportation systems [1]; law and order improved [1]; tourism encouraged [1]

6 Specific opportunities will depend on the city chosen. Points may include: redevelopment of old brownfield sites [1]; creation of new jobs in high tech industries [1]; transport improvements [1] (giving a specific example [1]); promotion of tourism [1]; regeneration projects [1] (giving a specific example [1])

7 Specific challenges will depend on the city chosen. Points may include: housing inequalities [1]; variations in educational achievement [1]; in-migration [1]; loss of secondary jobs in industrial areas [1], leading to unemployment [1]; pay inequality [1]; dereliction in old industrial areas [1]; pollution [1]

8 **Example points:** affordable house prices/rents [1]; shared housing [1]; natural lighting where possible [1]; passive heating wherever possible [1]; recycling [1]; triple-glazed windows [1]; renewable energy use [1]; water meters [1]

9 Examples must be given with both named strategies. Points may include: car sharing [1]; pedestrian-only areas [1]; street closures [1]; integrated public transport systems [1]; park-and-ride schemes [1]; bus lanes [1]; encouragement of cycling [1]; dedicated pedestrian walkways [1]

10 **Example points:** vertical farming / allotments / rooftop gardens [1]; locally grown produce [1] is sold locally [1]; removes need for long distance transportation [1]

11 **Example points:** aesthetically pleasing (looks nice) [1]; improves well-being / mental health [1]; provides shade on streets [1]; helps reduce air and noise pollution [1]; trees take in carbon dioxide [1]; provides oxygen from photosynthesis [1]; intercepts / soaks up rainfall/water [1]; helps to prevent flooding [1]

Pages 17–19: The Changing Economic World

1 **Example points:** data given is for the whole country [1]; some regions within a country might be more developed than others [1] and the figure given is only an average [1]

2 **Example measures:** life expectancy [1]; GNI [1]; GDP [1]; literacy rate [1]; birth rate [1]; death rate [1]; people per doctor [1]; calorie intake [1]; car ownership [1]; access to safe water [1]; infant mortality rate [1]

3 Population changes over time [1], which can be related to stages of development [1]

4 **Example points:** climatic extremes [1]; unproductive land for farming [1]; unreliable water supplies [1]; natural hazards [1]; limited natural resources [1]

5 **Example points:** disparity in wealth distribution [1]; disparities in health and medical services [1]; international migration [1]

6 **Example points:** appropriate simple technology [1] as the most up-to-date might not be suitable to the conditions [1]; specific example [1]

7 **Example points:** where raw material producers are paid a fair price for their goods [1] such as Fair Trade bananas/coffee/cotton [1]; allows farmers to plan ahead [1], knowing their income is secure [1]

8 A named LIC or NEE must be given. Points may include: can account for large percentage of GDP [1]; local cultures are preserved [1]; establishment of national parks protect wildlife and landscapes [1]; improvements to infrastructure for visitors [1] also benefits local people [1]; improved quality of life [1] as a result of increased levels of income [1]

9 A named example must be given. Points may include:
Advantages: improved levels of wages [1]; increased levels of female employment [1]; new skills [1]; increased levels of tax paid [1]; allows greater spend by government on infrastructure [1]; attracts other TNCs [1]
Disadvantages: national cultures can be undermined [1], by westernisation [1]; TNC may exert 'control' over government decision making [1]; environmental impacts / pollution [1]; excessive use of water [1]; TNC could move away at short notice [1]

10 A named country must be given. Points may include:
Short-term aid: to provide assistance following a natural disaster [1]; providing temporary housing [1]; medical support [1]; repair to infrastructure/roads/airport [1]
Long-term aid: major project such as building a dam/hospitals/airport [1]; developing the education/medical infrastructure [1]; developing natural resources [1]

11 **Example points:** traditionally, the north has heavy industry [1] and the south has services and high tech industries [1]; wages higher in the south [1]; unemployment higher in the north [1]; house prices higher in the south [1]; students have better exam results in the south [1]; life expectancy is longer in the south [1]

12 **Example points:** when primary (raw materials) and secondary (manufacturing) industries have declined [1], as a result of mechanisation and automation [1], exhaustion of local raw materials [1] and/or competition from overseas [1]

13 **Example points:** use of modern technology to reduce emissions [1]; named example [1]; desulphurisation [1]; tighter controls on water pollution [1]; restoration of damaged landscapes [1]; fines for pollution [1]; greater use of old brownfield sites for future development [1]

14 **Example points:** loss/closure of shops [1] and other local amenities such as primary school [1], doctor [1], church [1] and bus

services **[1]**; out-migration of younger people **[1]**; in-migration of second homers and weekenders **[1]**

15 Named examples must be given. Points may include:
Social benefits: people moving away from crowded urban areas **[1]**, such as the south-east **[1]**; less stressful journeys to work **[1]**; better national links **[1]**; easier access to airports **[1]**
Economic benefits: job creation **[1]**; increases people's access to job opportunities **[1]**; increased speed of business deliveries **[1]**; encourages business investment **[1]**; can lead to a positive multiplier effect **[1]**

Pages 20–23: The Challenge of Resource Management

1 The amount of carbon dioxide released into the atmosphere **[1]** as a result of the activities of a particular individual, organisation, or community. **[1]**

2 **Example points:** only 25% of our food is grown in the UK **[1]**, so we import from far afield **[1]**; demand is for crops 'out of season' in UK **[1]**, e.g. asparagus from Peru **[1]**

3 **Example points:** increasing affluence **[1]**; increasing population **[1]**; two-bathroom houses **[1]**; more appliances, e.g. washing machines **[1]**

4 **Example points:** highest demand in the driest part of the country **[1]**, i.e. London and the south-east **[1]**; least demand in the wettest part of the country **[1]**, i.e. the north and west **[1]**; water sent by pipeline and river **[1]**, e.g. Thirlmere to Manchester **[1]**; London faces significant problems in the future as it is so far from supplies **[1]**

5 Fossil fuels are running out **[1]**; less than 40% (2021) from renewables **[1]**, although increasing **[1]**; oil, gas etc. imported **[1]**; foreign supplies not guaranteed **[1]**; countries moving away from Russian supplies following the invasion of Ukraine in 2022 **[1]**

6 **Example points:** There are many different groups of countries: some, like those in the Middle East and North Africa, are arid and have shortages of water **[1]**. Others, like South Africa and India **[1]**, have poor infrastructure and are not able to maintain supplies to their population **[1]**. Other countries, like the UK and China **[1]**, have large population densities **[1]** and water in some areas may be in short supply due to high demand **[1]**.

7 Energy generated using resources that can be used over and over again **[1]**; examples include solar, wind, wave, tidal, hydro. **[2]**

8 **Example points:** noise **[1]**; increased traffic pollution **[1]**; risk of 'earthquakes' **[1]**; pollution from spillages **[1]**; environmental damage **[1]**; aesthetically damaging **[1]**; reduced house prices **[1]**

9 **Example points:**
Physical reasons: water availability **[1]**; high temperature **[1]**; poor soils **[1]**
Human reasons: population size **[1]**; skill levels **[1]**; lack of investment **[1]**; emphasis on producing export crops **[1]**

10 **Example points:** extreme climatic conditions **[1]**, such as drought **[1]** or other natural disaster example **[1]**; low yields **[1]**; environmental degradation **[1]**; soil erosion **[1]**; desertification **[1]**; social unrest / civil war **[1]**

11 The application of water to increase yields in areas where water is in short supply **[1]**

12 **Example points:** development of new crops that produce higher yields **[1]**, such as IR8 (a variety of high-yielding rice) in India **[1]**; development of crops that need fewer chemical inputs **[1]**; farmers able to afford to grow them **[1]**

13 Named example must be given. Points may include:
producers grow for their own use **[1]**; vegetables, fruit, bees, etc. **[1]**; surplus can be sold locally **[1]**; small-scale drip irrigation methods **[1]**; organic waste is recycled to enrich the soil **[1]**

14 Named examples must be given. Points may include:
water transfer systems **[1]**; desalinisation plants **[1]**; water conservation **[1]**, e.g. two-flush toilets **[1]**; use of greywater for toilets **[1]**; drought-resistant plants in gardens **[1]**; water butts used rather than hoses **[1]**

15 **Example points:** increased population means greater demand **[1]**; increased economic development leads to increased consumption **[1]**, both domestic and industrial **[1]**; affluence means higher living standards **[1]**, including more electrical items and gadgets **[1]**

16 **Example points:** intelligent building design **[1]**; insulation / double glazing to prevent heat loss **[1]**; energy-efficient devices **[1]**; auto switch-off features **[1]**; LED light bulbs **[1]**; cold-wash washing machines **[1]**; smart meters to control temperatures **[1]**; encouraging use of public transport rather than cars **[1]**

Page 24: Geographical Applications

1. Answers will vary. **[8]**

PRACTICE EXAM PAPERS

How exam papers are marked
Point marking
1, 2 and 3-mark questions are point marked. This means that if you give a correct answer, you attain a mark. For 2 and 3-mark questions, you can often earn a mark by developing your answer. For example, an answer of traffic congestion would earn 1 mark, but by linking it to longer journey times or delays to deliveries would earn an extra development mark.

Level marking
4, 6 and 9-mark questions are level marked. Each level has a description. The marker will read your answer and determine its level. Once they have assigned a level, they will then decide on a mark. The table below shows you some outline descriptors for each of the different levels.

Type of Question	Level 1 – Basic	Level 2 – Clear	Level 3 – Detailed
4 marks	Limited understanding. Limited/Unclear use of the figure. Limited application of knowledge and understanding.	Clear understanding. Effective use of the figure. Applies knowledge and understanding. Answers all aspects of the question.	
6 marks	Limited understanding. Limited application of knowledge and understanding. Limited use of the figure and/or a named example or case study.	Clear understanding. Some application of knowledge and understanding. Some use of the figure. Uses a named example or case study.	Detailed understanding. Thorough application of knowledge and understanding. Clearly uses the figure. Detailed use of a named example or case study. Answers all aspects of the question.
9 marks	Limited knowledge. Limited understanding. Limited application of knowledge and understanding. Limited use of the figure and/or a named example or case study.	Reasonable knowledge. Clear understanding. Reasonable application of knowledge and understanding. Some use of the figure. Some use of a named example or case study.	Detailed knowledge. Thorough understanding. Thorough application of knowledge and understanding. Accurate use of the figure. Detailed use of a named example or case study. Answers all aspects of the question. Draws a considered conclusion.

Assessment of spelling, punctuation, grammar and use of specialist terminology (SPaG)
Accuracy of spelling, punctuation, grammar and the use of specialist terminology will be assessed when indicated beside a 9-mark question. The cover sheet of each exam paper identifies which questions will be used. In each of these questions, 3 marks are allocated for SPaG as follows:
High performance – 3 marks
- spelling and punctuation is consistently accurate
- well-written with excellent use of grammar, so meaning is always clear
- a wide range of specialist terms are used appropriately

Intermediate performance – 2 marks
- spelling and punctuation is considerably accurate

- good use of grammar, so meaning is generally clear
- a good range of specialist terms are used appropriately

Threshold performance – 1 mark
- spelling and punctuation is reasonably accurate
- reasonable use of grammar (overall, any errors do not significantly hinder understanding of the answer)
- a limited range of specialist terms are used appropriately

No marks awarded – 0 marks
- no answer has been given
- the learner's response does not relate to the question
- spelling, punctuation and grammar does not reach the threshold level (errors mean that the answer cannot be properly understood)

For the purpose of this book, answers have been provided for 1, 2 and 3-mark questions so that you can mark your own answers. For level-marked 4, 6 and 9-mark questions, key points that should be addressed have been provided along with a model answer so that you can see what a good answer looks like. It is important to remember that you will only be awarded full marks in the exam if these points are communicated in a clear, accurate and well-developed way.

Pages 25–46
Paper 1: Living with the physical environment
Section A: The challenge of natural hazards
01.1   C **[1]**
01.2   Destructive plate margin / Convergent plate margin **[1]**
01.3   Your answer must refer to both places to show the contrast.
         Example:
         The UK is well within a tectonic plate / the Eurasian Plate, so shockwaves have further to travel / will be absorbed before affecting the UK **[1]** but three tectonic plates (may be named) meet at Japan **[1]**.
01.4   This question is level marked. Your answer must:
         - refer to Figure 2
         - cover both primary and secondary effects
         - include a range of ideas.
         You could include primary and secondary effects from either a volcano or earthquake event that you have studied.
         Example points:
         In Figure 2, ground shaking from the earthquake looks like it has damaged some buildings, particularly those to the right of the photograph. This is a primary effect. These homes may not be safe to live in, causing people to temporarily live somewhere else. This may cause stress to the residents. In the Eyjafjallajökull volcano eruption in Iceland, ash affected local farms and caused the loss of some livestock. Ash also caused the cancellation of thousands of European flights, leading to travellers being stranded across Europe and a loss of business to some companies. In Figure 2, it looks like the ground shaking has caused soil liquefaction. This is a secondary effect and has caused damage to roads, which may affect accessibility in the area. In Iceland, the publicity created by the volcano eruption has led to a long-term increase in tourism, which has created jobs and provided more income to some businesses.
01.5   B **[1]** and D **[1]**
01.6   Winds blow from areas of high pressure to areas of low pressure **[1]**
01.7   This question is level marked. Your answer must:
         - show a clear understanding of the causes of tropical storms
         - include several developed explanations.
         Example points:
         Tropical storms form over warm oceans between 5° and 20° north and south of the Equator. Sea surface temperature must be 27°C or more. The warm oceans cause air to rise; this is low pressure. As the air rises, it cools and condenses forming large, towering clouds leading to convectional rain. The Coriolis Force causes a tropical storm to rotate, anticlockwise in the northern hemisphere and clockwise in the southern hemisphere.
01.8   You could agree or disagree with the statement depending on the evidence you use.
         Example points:
         There is a link between maximum wind speed and the number of deaths. The two tropical storms with the highest maximum wind speeds caused the most deaths **[1]**. Hurricane Katrina with a wind speed of 175 mph caused 1200 deaths; this is four times the number of deaths caused by Superstorm Sandy, which had the lowest maximum wind speed at 110 mph **[1]**.

There is no clear link. Some of the storms with the highest wind speeds caused low number of deaths **[1]**. For example, Hurricane Michael had faster wind speeds than Superstorm Sandy, but Superstorm Sandy caused approximately six times more deaths **[1]**.
01.9   They may affect areas further away from the Equator / They may affect larger parts of the world **[1]**. They will become more intense / frequent **[1]**.
01.10   This question is level marked. There are also 3 marks for spelling, punctuation and grammar. Your answer must:
         - refer to Figure 5
         - cover both mitigation and adaptation strategies
         - include a range of well explained ideas
         - include ideas related to mitigation and adaptation from your own studies
         - use named examples
         - include a conclusion where you summarise your overall answer to the question.
         Example points:
         Figure 5 shows alternative energy production strategies, such as solar panels on house roofs and wind turbines. These are renewable energy sources and, unlike fossil fuels, don't release carbon dioxide into the atmosphere and therefore help mitigate against climate change. Many countries like the UK are trying to develop cleaner sources of energy. Another mitigation strategy shown is planting trees. Trees are carbon sinks. They take in carbon dioxide through photosynthesis, and this reduces the amount in the atmosphere, therefore helping combat climate change. Mitigation strategies are an important way of managing climate change as they stop the causes of climate change by reducing the amount of greenhouse gases in the atmosphere. In countries such as Bangladesh, they are adapting to the effects of climate change by building houses on stilts so that they are not affected by rising sea levels. This helps to reduce damage to people's property and possessions. Figure 5 shows farming in what looks like very dry conditions. In countries such as Burkina Faso, they are growing crops that are drought resistant. This enables farmers to be able to continue to provide food and earn a living despite rising temperatures and longer periods of drought caused by climate change. In places such as Cape Town in South Africa, they are investing in water supply strategies like desalination to ensure water security even when there are extreme droughts. In conclusion, both adaptation and mitigation strategies are needed to manage climate change. Adaptation enables us to continue to live within a changing environment, whilst mitigation strategies are needed to limit the amount that the climate changes.

Section B: The living world
02.1   B **[1]**
02.2   B **[1]**
02.3   A community of plants and animals / A community of living and non-living elements **[1]**
02.4   Decomposers help to return nutrients/energy to the soil / Decomposers break down dead plants and animals or excreted material **[1]**
02.5   **Any one from:** Insect; rabbit; mouse; red deer; bird; hare **[1]**
02.6   Fewer red deer means less demand on plants **[1]**, so more of those are available to other primary consumers, e.g. rabbits, insects **[1]**. The numbers of other primary consumers might increase as a result of more food being available **[1]**. Less food available for eagles **[1]** because they are the only predator of red deer **[1]**. Eagle numbers might decrease **[1]**. Eagles would hunt more rabbits, mice or hares **[1]**, whose numbers could decline **[1]**.
02.7   Rainfall occurs all year round **[1]**. There are seasonal patterns of rainfall (named months may be given) **[1]**. The highest rainfall, 295 mm, occurs in March **[1]**. The lowest rainfall, 55 mm, occurs in August **[1]**. The range in rainfall is 240 mm **[1]**.
02.8   The sun is much higher in the sky for most of the year **[1]** / The sun's rays have a relatively short distance to travel **[1]** / The sun's rays are more concentrated **[1]**
02.9   This question is level marked. Your answer must:
         - refer to Figure 9
         - cover both economic and environmental impacts
         - include a range of ideas.

Example points:
I am against the development of wind energy in the Scottish Highlands.

One reason is that wind energy in not reliable as it depends on there being wind. In Resource 3, it mentions that, in January 2017, the 'wind died for a week'. This would result in energy insecurity and could cause problems for people and businesses. Investment in nuclear power would be better as this type of energy is more reliable.

Resources 7 and 8 refer to some of the negative environmental impacts of wind energy generation. The Scottish Highlands is an area of natural beauty. I think that this should be protected. Developing wind turbines would change the appearance of the landscape negatively. This may deter visitors, affecting tourism in the area. This could lead to a loss of trade for businesses relying on tourism, such as hotels, which may lead to job losses. It could also have a negative effect on the price of local people's houses. Building wind turbines may disturb habitats, which could drive wildlife out of the area. Furthermore, as mentioned in Figure 8, the construction of wind turbines can affect bird flightpaths and could even kill some birds. This would affect food chains.

I am not completely against the development of wind energy. It does bring benefits through job creation and, once built, this type of energy does not create carbon emissions and this will help to mitigate climate change. As Resource 6 shows, the amount of electricity being created by wind power is increasing. However, I don't think these turbines should be built in areas of natural beauty. They could be built in areas that have already been developed; for example, near to motorways or industry. Overall, I am against building wind turbines in the Scottish Highlands. However, I am not against developing wind energy. I think that the UK should develop wind energy, along with other sources such as solar and nuclear. This will create a diverse energy mix, ensuring future energy security.

Section B: Fieldwork

04.1 C **[1]**

04.2 It might be difficult to get to **[1]** because the land may be very steep **[1]**. The land might be private **[1]** so not accessible to the public **[1]**. There may be animals in the area **[1]**, so it would be dangerous **[1]**.

04.3 Bar chart **[1]**
Easy to compare different locations **[1]**; easy to see differences between different locations **[1]**; consists of discrete/discontinuous data **[1]**; gives an accurate number for each site **[1]**

04.4 Advantage – considers all the data **[1]**; you do not need to arrange the data as you do when you calculate the median **[1]**
Disadvantage – very small or large values can affect the mean **[1]**; anomalous values can distort the mean **[1]**; it may not give a true central value of the data **[1]**

04.5 There is a positive correlation **[1]**. As distance down the valley increases, the mean length of stones increases **[1]**. The mean length of stone increases by nearly three times between Location A and Location E **[1]**.

04.6 Correct completion of the graph (i.e. 'Good' 22 to 40 and 'Very good' 40 to 50). Must use shading as indicated on the key. **[1]**

04.7 56% **[1]**

04.8 It is easy to see differences in the opinions of people about the quality of the two footpaths **[1]**; it provides accurate numbers **[1]**; it is an appropriate method for discrete/discontinuous data **[1]**.

04.9 This question is level marked. Your answer must:
- refer to Figure 9
- identify similarities and differences between the width of the two footpaths.

Example points:
Figure 9 shows that both footpaths are narrower at the start and at around 900 metres along the footpath. Both footpaths are at their widest at about 700 metres. Footpath A is always wider than Footpath B. Footpath A starts to get wider after 200 metres whilst Footpath B starts to get wider after about 300 metres. Footpath A narrows at 900 metres whereas Footpath B narrows at 800 metres.

05.1 A risk is drowning in a river / sea **[1]**. This risk can be reduced by checking river levels / tide times **[1]** and by staying back from the water's edge **[1]**. A risk is tripping on uneven ground **[1]**. This risk can be reduced by wearing appropriate footwear **[1]**

and staying on designated footpaths **[1]**. A risk is cliff rock fall **[1]**. This risk can be reduced by staying back from the cliffs **[1]** and wearing protective headgear **[1]**. A risk is severe sunburn / sunstroke **[1]**. This risk can be reduced by wearing a hat **[1]** and regular application of sun cream **[1]**.

05.2 This question is level marked. Your answer must:
- refer to the method used in a physical geography fieldwork investigation
- refer to specific data collection methods that you used
- recognise the advantages and disadvantages of your data collection method(s)
- include a range of well explained ideas.

Example points:
One method of data collection I used was river velocity measurements. I carried out these measurements at three systematically-spaced locations along the River X in Cheshire. I used a flow meter; this should mean the measurements are accurate. At each location I measured the velocity three times and then calculated the mean. This reduced the influence of anomalies, making the data more reliable. I made sure that I measured the speed of the river where I thought it was fastest flowing, ensuring a consistent approach at all three sites. By collecting data at three sites downstream, I was able to answer my hypothesis that river velocity increased downstream. A weakness of my data collection method was that, by only doing three readings, an anomalous reading could have a significant impact on my data. It was hard to identify where the fastest flowing water was occurring, so I may not have carried out my measurements in the correct location. Furthermore, three sampling sites is a small sample size so my data may not be a true representation of the river. The data was also only collected at one time in the year after several weeks of little rain, which may not be representative of other times.

05.3 **Example points:** It was within walking distance of the school **[1]**, so it did not take long to get there **[1]**. There were no busy roads nearby **[1]**, so it was a safe area in which to collect data **[1]**. There was a range of survey points available **[1]** with enough variation that we could look at how environmental quality changes over distance **[1]**.

05.4 This question is level marked. There are also 3 marks for spelling, punctuation and grammar. Your answer must:
- refer to either your physical or human geography fieldwork investigation
- evaluate the degree to which the data collected allowed you to reach valid conclusions
- refer to specific evidence
- include a range of well explained ideas
- include a conclusion where you summarise your overall answer to the question.

Example points:
The title of the fieldwork investigation was 'Does quality of life increase with distance from the town centre?'.

I collected housing and environmental quality data at 10 systematically-spaced locations on a transect going through the town of X. The sample size was large enough that I could present the data in the form of scatter graphs. These clearly showed a positive correlation, meaning that environmental quality and housing quality increased with distance from the town centre. There were some anomalies, so I could not completely agree with my hypothesis.

Whilst the data I collected enabled me to reach conclusions, there are limitations with my data which may mean that these conclusions are not completely valid. One limitation of my study is that I only studied one transect within the town. Other areas may not demonstrate the same characteristics, meaning that my data may not be representative of the whole town. Many of the surveys I carried out, such as the environmental quality survey, were subjective and just based on my opinion. This means that the data may be biased and therefore the conclusions I have drawn may not be entirely valid. Furthermore, some of the secondary data I used, census data, was from 2011 and therefore out of date. It may no longer represent the characteristics of the people that live in the area today.

In conclusion, the data I collected did enable me to reach conclusions, although they may not be entirely accurate.

Acknowledgements

The authors and publisher are grateful to the copyright holders for permission to use quoted materials and images.
Page 6: © Citynoise at English Wikipedia/wikipedia.org
Page 12: © Alan Curtis / Alamy Stock Photo
Page 27: © Aflo Co. Ltd. / Alamy Stock Photo
Page 31: © Nick Hanna / Alamy Stock Photo; © ZUMA Press, Inc. / Alamy Stock Photo; © John Peter Photography / Alamy Stock Photo
Page 36: © Paralaxis / Alamy Stock Photo
Page 38: © Crown copyright and database rights (2022) Ordnance Survey (AC0000808974)
Page 40: © David Thompson / Alamy Stock Photo; geogphotos / Alamy Stock Photo
Page 41: © Crown copyright and database rights (2022) Ordnance Survey (AC0000808974)
Page 43: © Ian Goodrick / Alamy Stock Photo; © Paul Glendell / Alamy Stock Photo
Page 44: © Crown copyright and database rights (2022) Ordnance Survey (AC0000808974)
Page 46: © Neil McAllister / Alamy Stock Photo; © Vivienne Crow / Alamy Stock Photo
Page 50: © Paul Lovelace / Alamy Stock Photo
Page 51: © Dave Ellison / Alamy Stock Photo; © Mark Waugh / Alamy Stock Photo
Page 62: © Orapin Joyphuem / Alamy Stock Photo; © Erberto Zani / Alamy Stock Photo
Page 65: © Shine-a-light / Alamy Stock Photo; © Andrew Woodacre / Alamy Stock Photo
Page 68: © aerial-photos.com / Alamy Stock Photo; © Paul Glendell / Alamy Stock Photo
Page 73, 82 © Bjmullan/commons.wikimedia.org
Page 81: US Energy Information Administration, based on Digest of UK Energy Statistics and National Statistics: Energy Trends
Page 86: Map: Crown copyright 2021 / licensed under terms of the Open Government Licence

All other images Shutterstock.com and HarperCollins*Publishers*

Every effort has been made to trace copyright holders and obtain their permission for the use of copyright material. The authors and publisher will gladly receive information enabling them to rectify any error or omission in subsequent editions. All facts are correct at time of going to press.

Published by Collins
An imprint of HarperCollins*Publishers* Ltd
1 London Bridge Street
London SE1 9GF

HarperCollins*Publishers*
Macken House, 39/40 Mayor Street Upper, Dublin 1, DO1 C9W8, Ireland

© HarperCollins*Publishers* Limited 2022

ISBN 9780008535056

First published 2022

10 9 8 7 6 5 4 3 2

All rights reserved. No part of this publication may be reproduced, stored in a retrieval system, or transmitted, in any form or by any means, electronic, mechanical, photocopying, recording or otherwise, without the prior permission of Collins.

British Library Cataloguing in Publication Data.

A CIP record of this book is available from the British Library.

Publisher: Clare Souza
Authors: Paul Berry, Brendan Conway, Janet Hutson, Dan Major, Tony Grundy, Robert Morris and Iain Palôt
Videos: Eleanor Barker
Project Management and Editorial: Richard Toms
Cover Design: Sarah Duxbury and Kevin Robbins
Inside Concept Design: Sarah Duxbury and Paul Oates
Text Design and Layout: Ian Wrigley
Artwork: Ian Wrigley
Production: Lyndsey Rogers
Printed in the UK, by Ashford Colour Press Ltd.

MIX
Paper | Supporting responsible forestry
FSC™ C007454
www.fsc.org

This book contains FSC™ certified paper and other controlled sources to ensure responsible forest management.

For more information visit: www.harpercollins.co.uk/green